ATMOSPHERIC HAZARDS - CASE STUDIES IN MODELING, COMMUNICATION, AND SOCIETAL IMPACTS

Edited by **Jill S. M. Coleman**

Atmospheric Hazards - Case Studies in Modeling, Communication, and Societal Impacts

http://dx.doi.org/10.5772/61419
Edited by Jill S. M. Coleman

Contributors

Ronald Holle, Mary Ann Cooper, César Augusto Aguirre, Armando Benito Brizuela, Nicolas Freychet, Huang-Hsiung Hsu, Chi-Hua Wu, John A. Knox, Brendan Mazanec, Emily Sullivan, Spencer Hall, Jared Rackley, Kevin Law, Robert Schwartz

CBS, Edition **2017**

Published by InTech

Janeza Trdine 9, 51000 Rijeka, Croatia

Publishing Process Manager Romina Rovan

Technical Editor SPi Global

Cover Designer InTech Design Team

Additional hard copies can be obtained from orders@intechopen.com

Atmospheric Hazards - Case Studies in Modeling, Communication, and Societal Impacts, Edited by Jill S. M. Coleman
p. cm.
Print ISBN 978-953-51-2515-0
Online ISBN 978-953-51-2516-7

INTECH

open science | open minds

Section 1

Part I: Atmospheric Hazards and Societal Implications

Chapter 1

Lightning Occurrence and Social Vulnerability

Ronald L. Holle and Mary Ann Cooper

Additional information is available at the end of the chapter

http://dx.doi.org/10.5772/63001

Abstract

The occurrence of lightning in time and space around the world is well known. Lightning fatalities and injuries are well delineated in the United States; however, there is much less information about lightning impacts on people in the developing world. It is estimated that between 6000 and 24,000 people are killed globally per year, and 10 times as many are injured. The fatality rate per capita has become very low in the developed countries during the past century due to the availability of lightning-safe structures and vehicles, less labor-intensive agriculture, and other factors, but this reduction has not occurred where people continue to work and live in lightning-unsafe situations. Lightning safety advice often mistakenly expects that the direct strike is most common, but ground current, direct contact, side flash, and upward streamers are much more frequent mechanisms. In developed countries, the injury:death ratio is approximately 10:1, meaning that 90% survive but may have permanent disabling injuries. The proximate cause of death is cardiac arrest and anoxic brain injury at the time of the lightning strike, and, at this time, the damage from a lightning strike cannot be reversed or decreased in survivors. Lightning vulnerability in many developing countries continues to be a major issue due to widespread exposure during labor-intensive agriculture during the day when thunderstorms are the most frequent and while occupying lightning-unsafe dwellings at night.

Keywords: lightning, lightning strikes, lightning fatalities, lightning safety, lightning occurrence

1. Introduction

More lightning occurs in clouds than that strikes the ground; and as the cloud moves above the earth, opposite charges are induced on the surface of the earth and on objects on the ground under the cloud. Upward streamers, not usually visible from these objects, will reach up and attempt to connect with the downward-moving lightning channel. A cloud-to-ground flash has one or more return strokes. When you see lightning flickering, those are return strokes within the same flash. There are about four strokes per flash, when averaged over a large sample size. The first stroke comes faintly to ground and contacts the surface of the earth. The stroke may attach to a tree, open land, the ocean, or other objects. Then the light fills the channel going upward. The next stroke in the flash will likely come down from the cloud to the ground in the same channel as the first stroke, and so on. Occasionally, one of the subsequent strokes will stray from the pre-existing channel and come to ground one or two kilometers away.

One of the important features of this well-known mechanism of lightning strikes is the search radius. The cloud-to-ground channel makes its way downward in step leaders of about 50-m lengths (**Figure 1**). At the lower tip of each step, the channel searches for a feature that makes a convenient connection to ground. Higher up in the cloud, there is nothing to strike, so the step leader keeps coming toward the ground as each step attempts to reach out and connect to the ground. Branching occurs that is generally downward. Only when the lowest tip of the channel is 30–50 m from the surface of the earth does it 'decide' what to strike. This last step leader, not usually visible from a distance, is nearly vertical and connects one of the upward streamers emanating from objects such as trees, poles, and sometimes open water. The most likely connections will be made to objects that are tall, isolated, and pointed. But if that object is more than 30–50 m away from the lowest tip of the downward leader, the lightning channel may come all the way to ground, close to a tall building or tower.

Figure 1. Cloud-to-ground lightning photographed from Oro Valley, Arizona, USA (©R. Holle).

2. Lightning occurrence

The location, time of year and day, and number of lightning occurrences throughout the world are known quite well. **Figure 2** shows the latest 10-year map of the United States cloud-to-ground flash density per square kilometer. With respect to time of day, about two-thirds of

http://dx.doi.org/10.5772/63001

lightning occurs during the afternoon from noon to 1800 local time. Cloud-to-ground lightning is the most frequent along the coasts of Florida and the Gulf of Mexico, and generally decreases to the north and west as the number of days with substantial low-level moisture decreases in frequency. Large variations in lightning frequency also occur in the western states where large elevation changes affect thunderstorm formation.

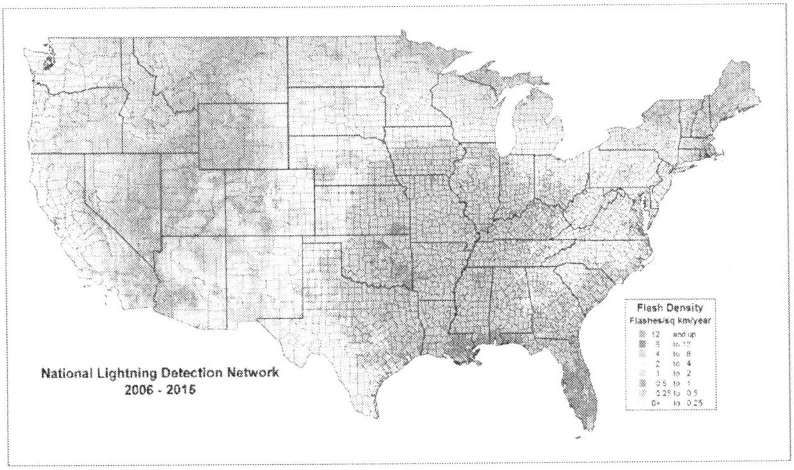

Figure 2. Cloud-to-ground flash density per square kilometer per year over the contiguous United States from the National Lightning Detection Network from 2006 through 2015 (Courtesy Vaisala, Inc.).

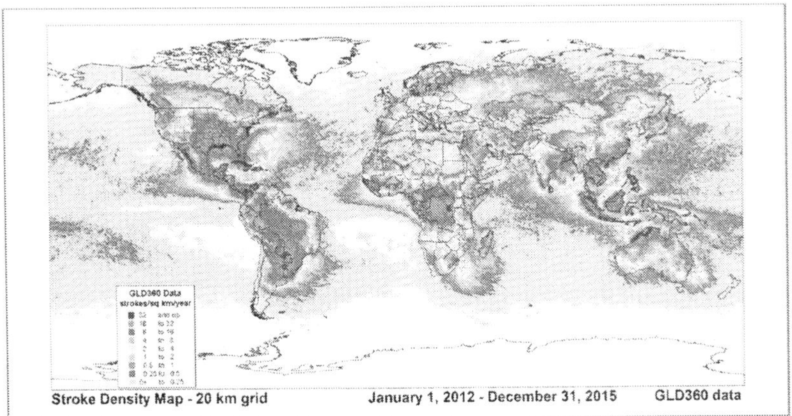

Figure 3. Lightning stroke density per square kilometer per year over the world from the Global Lightning Dataset GLD360 from 2012 through 2015 (Courtesy Vaisala, Inc.).

Figure 3 shows the global occurrence of lightning strokes. The highest densities are over tropical and subtropical coastlines, near large elevation changes, and east coasts at middle latitudes. The lowest lightning frequency is over oceans, the polar regions, and west coasts at middle latitudes. The dominance of land lightning is due to daytime heating that can produce updrafts rising to altitudes where temperatures are colder than freezing, the altitudes where lightning initiates. Since much of outdoor human activity occurs during daytime, the juxtaposition of lightning with people can be expected to result in fatalities and injuries.

Nearly everywhere in the world has a season with more lightning than others. In the middle latitudes about two-thirds of lightning occurs during meteorological summer (June, July, and August in the Northern Hemisphere and December, January, and February in the Southern Hemisphere). In the Tropics, the passage of the equatorial trough or the summer monsoon strongly affects the frequency of lightning. The equatorial trough (also known as the Intertropical Convergence Zone, or ITCZ) is a somewhat continuous east-west area of rain and thunderstorms that stays within about 20 degrees latitude of the equator, shifting northward (southward) during the Northern (Southern) Hemisphere summer. For locations near the equator, the equatorial trough crosses locations twice a year which can result in two rainy seasons whereas more subtropical locations have only a single rainy season. The Asian Monsoon can be considered to be a variation of the equatorial trough that is strongly affected by the large land mass of Asia and the Himalayas. The equatorial trough generally moves northward in the Northern Hemisphere summer and brings tropical moisture from the south, often resulting in higher thunderstorm and lightning frequencies, and then reverses to send dry air flowing from the north during the Northern Hemisphere winter when lightning occurrence is much less likely.

3. Lightning fatalities and injuries

3.1. United States fatalities and injuries

According to the U.S. National Weather Service (NWS), lightning fatalities have averaged 32 per year over the past decade. **Figure 4** shows the latest decade of available fatality data with the top panel showing the number of fatalities by state while the lower panel indicates the fatality rate per million people. The fatality rate turns out to be rather different from the actual number of fatalities, and indicates how the number of lightning flashes and the number of fatalities are related, but not as directly as may be expected. States with the highest fatality rates in the western United States appear to indicate a region with drier air and lower rainfall rates that give the perception that lightning is not as much of a threat as when it is raining heavily in other locations in the United States.

http://dx.doi.org/10.5772/63001

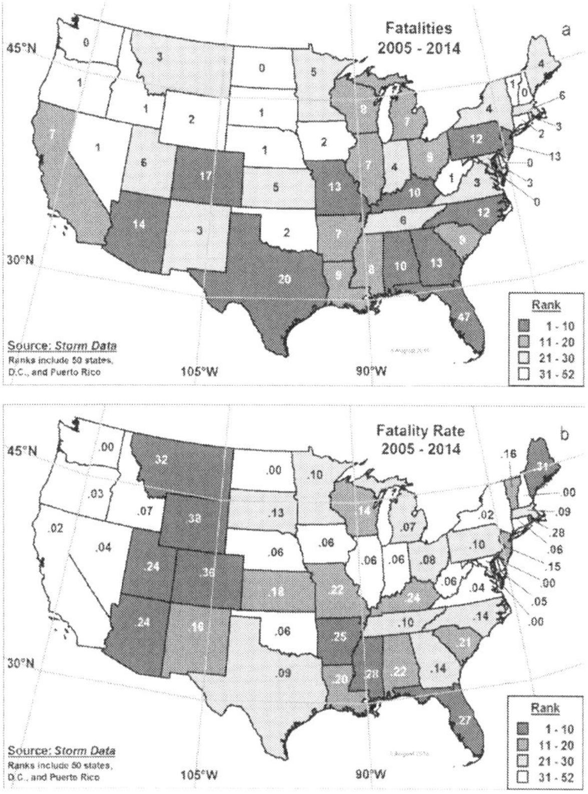

Figure 4. Ranks of lightning fatalities per state (a) and fatality rate per million people (b) from 2005 through 2014 based on the United States National Weather Service website [2].

The necessary ingredients of lightning frequency and population density in the United States have been combined in a recent study that is able to replicate the primary locations of observed lightning fatalities [1]. This approach shows the concentration of fatalities in urban areas that have moderate to high lightning frequencies. Such a study has not yet been attempted in other countries where the lightning risk is very different due to such factors as the availability of lightning-safe buildings and vehicles, agricultural participation, and related societal differences.

Underreporting of lightning casualties has been a problem in past years in the United States, but much less in the past couple of decades. A primary reason for underreporting is that about 90% of all lightning fatalities and injuries are to one person at a time, which leads to the tendency for such events to be reported less often in the media than multiple-casualty events. With the inception of the Lightning Safety Awareness Team, a multidisciplinary group sponsored by the National Weather Service, it is probable that every fatality is documented [2], but an estimate is that only about 70% of the injuries reach the reporting system that is maintained by all NWS offices across the country. An additional issue is that underreporting may seem to be occurring when there is actually a definition issue. The National Weather Service in its *Storm Data* publication does not include secondary casualties due to lightning. For example, a house caught on fire at night due to lightning that results in a fatality or injury is not counted as a lightning impact, since the primary cause is coded as fire not lightning. A useful method for estimating injury underreporting is the ratio of injuries to deaths. An intensive study over three full years across the state of Colorado showed that about ten injuries occur per fatality [3]. When the ratio is less than ten, an indication exists that not enough injuries have been reported and documented. The 10:1 injury:death ratio is assumed to apply to the United States and more developed countries with their similar socio-economic infrastructures, providing widespread availability of lightning-safe buildings and vehicles.

The lightning fatality rate per million people in the United States has dropped by more than two orders of magnitude since 1900 (**Figure 5**). Similar trends have been observed in many more developed countries of the world with published lightning fatality rates during the same period [4]. Also shown is the rural percentage of the population, which decreased from 60% in 1900 to fewer than 20% at present. The United States population not only transitioned out of a mainly labor-intensive agricultural society a century ago, but also moved into more substantial home and workplace buildings with grounded wiring and plumbing, together with the ready availability of fully enclosed metal-topped vehicles, better medical care, and greatly improved meteorological information about thunderstorms. All of these factors are present in more developed countries that have lower lightning death rates.

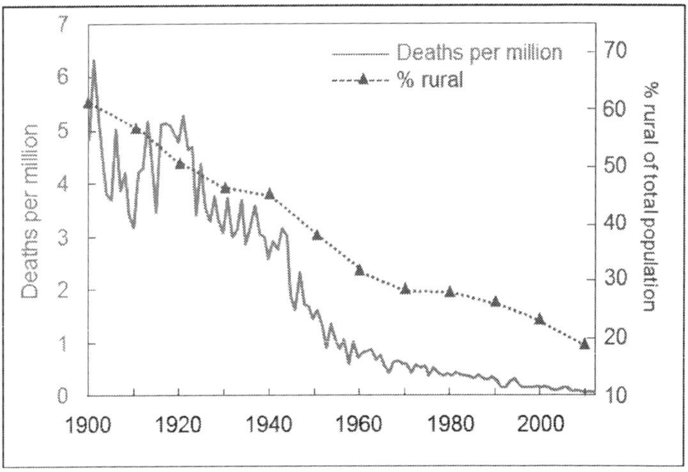

Figure 5. Solid red line: United States lightning deaths per million people from 1900 to 2013. Dashed blue line: percent rural population.

Figure 6 graphs United States lightning fatality types one hundred years apart, comparing the 1890s (top) with the 2005–2014 period (bottom). During the late nineteenth century, being indoors was the most common situation for lightning fatalities, while agriculture and outdoors were also a large component of the events. During the decade starting in 2005, indoor fatalities have become quite rare, agriculture events have greatly reduced from a century ago as farming became mechanized, and lightning-safe buildings and vehicles became common. Recreation and sports situations are relatively more frequent now than earlier. However, the overstated scenario of golf fatalities has been exceeded substantially in recent years by hiking, climbing, boating, and other water-related activities [2].

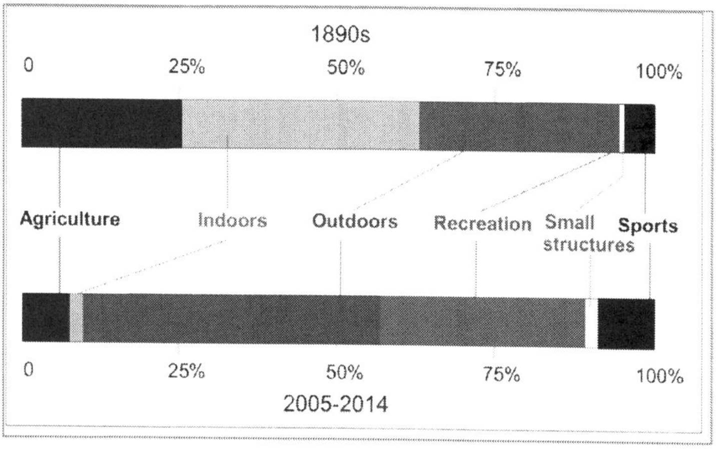

Figure 6. Comparison of the percentage of types of the United States lightning fatalities in the 1890s versus 2005 through 2014 [20].

3.2. Global fatalities and injuries

Worldwide, **Figure 7** shows information on lightning fatality rates that has been collected in a number of countries during the last quarter century; nevertheless, significant gaps exist in our knowledge of the absolute numbers [4]. Studies have estimated the number of deaths per year attributed to lightning globally anywhere from a few thousand to 24,000 [5–7]; however, much uncertainty exists due to the limited sample size (**Figure 7**). One country of particular interest is Malawi with a very high rate of 84 lightning deaths per million people per year that far exceeds the rate in any other country [4]. The 1008 annual fatalities for this small but populous country may represent a very complete data collection method that is what adjacent countries in the region should report, so very large numbers of lightning fatalities may be actually occurring but we do not know them. If the "10:1 (injury:death) rule" is appropriate for developing countries, then lightning is causing a large number of deaths and injuries worldwide, regardless of our inability to state the actual numbers.

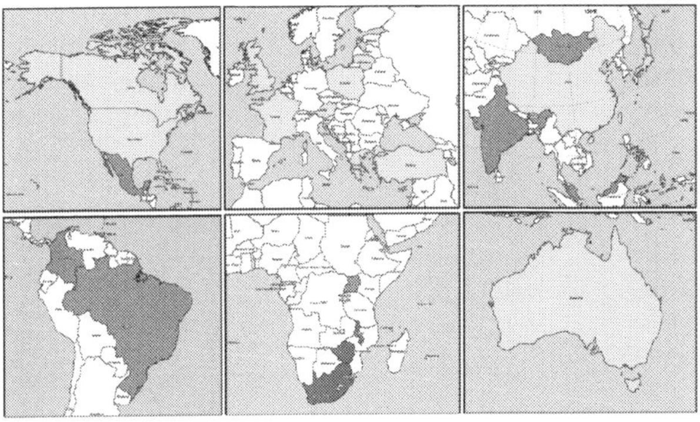

Figure 7. Lightning fatality rate per million people per year by continent. Red shading indicates rate >5.0 fatalities per million per year, orange is 0.6 to 5.0, and yellow is 0.5 or less. White indicates no national summaries have been published for datasets ending in 1979 or later (Updated with permission from Holle [4]).

Figure 8. Many forms of labor-intensive work are far from lightning safety such as (a) fishermen in open boats on Lake Victoria, the second largest lake in the world, and (b) fields in Nepal (courtesy ACLENet and M.A. Cooper).

http://dx.doi.org/10.5772/63001

In contrast with the United States, many populous less-developed countries have as much as 90% of the population living and working in lightning-unsafe locations and situations. During the day, many people are involved in labor-intensive agriculture, fishing in open boats, walking to market, or inside schools without recourse to safety in an appropriate building or vehicle [8] (**Figures 8** and **9**). At night, people live in lightning-unsafe dwellings without adequate wiring, plumbing, or metal structural components that can carry lightning into the ground without affecting individuals inside [9] (**Figure 10**).

Figure 9. Most forms of transportation or going to market in developing countries are not lightning safe such as (a) taxi in India, (b) ox cart in India, and (c) bota-bota taxi in Nepal (©M.A. Cooper).

Figure 10. An estimated 90% of sub-Saharan buildings and housing are not lightning-safe, such as (a) farming settlement with thatch and sheet metal roofs on mud brick walls in Zambia, (b,c) combination shops and homes in India (©M.A. Cooper).

In the developing world, the injury:death ratio is expected to be lower since fewer lightning-safe locations are available, resulting in a higher proportion of deaths. In addition, more people die per lightning event in the developing world than in the United States, particularly in agriculture and school events. For developing countries, counting both "primary" and "secondary" causes may be preferable until fatality rates decrease and the availability of safe locations increases.

In more developed countries, the dominant profile of lightning casualties is the young male. Risk-taking in recreation, workplaces, and organized sports tends to be dominated by this group of males between about 15 and 30 years old. However, in lesser developed countries, the distribution is substantially more equal between female and male, and the ages are much more disperse, both because of more widespread exposure. A recent study of labor-intensive agriculture in mainly India and Bangladesh shows that 47% of the fatalities and injuries were females as they work during the daytime when thunderstorms are most frequent [8]. The lack of lightning-safe dwellings, schools, and workplaces means that all ages and both genders are equally vulnerable at all times.

3.3. Damage to property and indirect impacts

In developed countries, precautions against the effects of lightning to property, electronics, and utility lines are usually routine and subject to building codes using well-accepted practices for public buildings such as churches, hospitals, and schools. In developing countries, the effects of lightning damage to property can have not only direct effects on the structure but also indirect economic effects. The impacts include food spoilage from lack of refrigeration after electrical failure, electrical parts and repairs are unaffordable or not available for days, hospitals are without power, and databases and expensive, irreplaceable electronics are damaged. These adverse effects can occur in countries that are already struggling with other pressing issues such as drought, HIV, underemployment, or civil strife. Individual families can suffer not only when one or more of their members is injured or killed by lightning but also when their livestock, often the major measure of wealth, are killed en masse by lightning (**Figure 11**).

Figure 11. Cattle killed by a cloud-to-ground lightning strike in South Africa (©I. Jandrell).

more frequent along coastlines and near large topographic features in the Tropics and subtropics. However, due to socio-economic factors, the number of fatalities and injuries is not as closely related to lightning frequency as might be expected. In the United States and other developed countries, the fatality rate per capita is more than two orders of magnitude lower than a century ago. This rate decrease is attributable to a shift from rural to urban settings where fewer people are involved in labor-intensive agriculture. Many lightning injuries and deaths in the developed world are from leisure activities, often related to various types of activities in the vicinity of water bodies [2]. At least as important has been improvements to the quality of buildings that are usually safe from lightning due to grounding according to codes that provide safety to people inside. In addition, the widespread availability of fully enclosed metal-topped vehicles provides mobile lightning-safe locations almost everywhere. Additional factors include better medical treatment and improved understanding of thunderstorms and their associated lightning threat.

In contrast, many developing countries continue to have very high per-capita fatality and injury rates. A large portion of their population is involved in labor-intensive agriculture during the daytime when thunderstorms are more common. No lightning-safe locations are typically available to these people, often nearly equally male and female, while working in the fields. In addition, dwellings occupied outside of working hours often are not lightning-safe due to their construction of mud brick and thatch or sheet metal roofing. As a result, an estimate is that as many as 24,000 deaths and 240,000 injuries occur from lightning globally every year [7], almost entirely in the less-developed countries of the world. These figures remain elusive, due to sporadic data gathering in the most lightning-vulnerable locations where data collection will be slow to improve in the near future such as equatorial countries of Africa and Asia. Injury prevention is far better than taking care of people after they are injured; both scenarios are relatively easier in developed countries. Unfortunately, the infrastructure and housing in developing countries precludes easy answers like "When Thunder Roars, Go Indoors", and lightning remains a substantial threat to entire villages, schools, and populations.

Author details

Ronald L. Holle[1*] and Mary Ann Cooper[2]

*Address all correspondence to: rholle@earthlink.net

1 Holle Meteorology & Photography, Oro Valley, Arizona, USA

2 Department of Emergency Medicine, University of Illinois at Chicago, Chicago, Illinois, USA

References

[1] Roeder WP, Cummins BH, Cummins KL, Holle RL, Ashley WS. Lightning fatality risk map of the contiguous United States. Natural Hazards, 2015;79:1681-1692. DOI 10.1007/s11069-014-1254-9.

[2] National Oceanic and Atmospheric Administration, National Weather Service, www.lightningsafety.noaa.gov, Victims/Medical, Analysis of Recent Lighting Fatalities.

[3] Cherington M, Walker J, Boyson M, Glancy R, Hedegaard H, Clark S. Closing the gap on the actual numbers of lightning casualties and deaths. In: Preprints of the 11th Conference on Applied Climatology, 10–14 January 1999; Dallas, Texas, American Meteorological Society; 1999. pp. 379-380.

[4] Holle RL. The number of documented global lightning fatalities. In: Preprints of the 6th International Lightning Meteorology Conference, 18-21 April 2016; San Diego, California, Vaisala; 2016, 4 pp.

[5] Gomes R, Ab Kadir MZA. A theoretical approach to estimate the annual lightning hazards on human beings. Atmospheric Research, 2011;101:719-725. doi:10.1016/j.atmosres.2011.04.020.

[6] Cardoso I, Pinto Jr. O, Pinto IRCA, Holle RL. A new approach to estimate the annual number of global lightning fatalities. In: Preprints of the 14th International Conference on Atmospheric Electricity (ICAE), 8–11 August 2011; Rio de Janeiro, Brazil, 2011. 4 pp.

[7] Holle RL, López RE. A comparison of current lightning death rates in the U.S. with other locations and times. In: Preprints of the International Conference on Lightning and Static Electricity, 16–19 September 2003; Blackpool, England, Royal Aeronautical Society, paper 103-34 KMS; 2003. 7 pp.

[8] Holle RL. Lightning-caused deaths and injuries related to agriculture. In: Preprints of the 6th International Lightning Meteorology Conference, 18–21 April 2016; San Diego, California, Vaisala; 2016.

[9] Holle RL. Lightning-caused casualties in and near dwellings and other buildings. In: Preprints of the International Lightning Meteorology Conference, 21–22 April 2010; Orlando, FL, Vaisala; 2010. 19 pp.

[10] Blumenthal R, West NJ. Investigating the risk of lightning's pressure blast wave. South African Journal of Science, 2015;111(3/4): 5; article 2014-0187.

[11] Holle RL. Lightning-caused deaths and injuries In the vicinity of trees. In: Preprints of the 6th Conference on the Meteorological Applications of Lightning Data, 6–10 January 2013; Austin, Texas, American Meteorological Society; 2013. 8 pp.

[12] Cooper MA, Holle RL. Casualties from lightning involving motorcycles. In: Preprints of the International Conference on Lightning and Static Electricity, 28–30 August 2007; Paris, France; 2007. paper Ic07/PPRKM02.

[13] Cooper MA, Holle RL, Andrews CJ, Blumenthal R. Lightning injuries. In: Auerbach P, editor. Wilderness Medicine, 6th ed., Elsevier Mosby; Philadelphia, PA. 2012. pp. 60-101.

[14] Cooper MA. Whether the medical aspects of lightning injury are different in developing countries. In: Preprints of the 32nd International Conference on Lightning Protection, 2–7 September 2012; Vienna, Austria, DOI:978-1-4673-1897-6/12/.

[15] Cooper MA. Disability, not death, is the main problem with lightning injury. National Weather Digest, 2001;25:43-47.

[16] Cooper MA, Marshburn S. Lightning strike and electric shock survivors, International. NeuroRehabilitation, 2005;20:43-37.

[17] Mulder MB, Msalu L, Caro T, Salerno J. Remarkable rates of lightning strike mortality in Malawi. PLoS One, 2012;7(1). DOI:10.1371/journal.pone.0029281.

[18] Roeder WP, Holle RL, Cooper MA, Hodanish S. Communicating lightning safety effectively. In: Preprints of the 5th Conference on the Meteorological Applications of Lightning Data, 23–27 January 2011; Seattle, Washington, American Meteorological Society; 2011. 18 pp.

[19] Gomes R, Kithil R, Ahmed M. Developing a lightning awareness program model for third world based on American-South Asian experience. In: Preprints of the 28th International Conference on Lightning Protection, 17–21 September 2006; Kanazawa, Japan; pp. 1240-1243.

[20] López RE, Holle RL. Changes in the number of lightning deaths in the United States during the twentieth century. Journal of Climate, 1998;11:2070-2077.

Chapter 2

Analysis of the Twitter Response to Superstorm Sandy: Public Perceptions, Misconceptions, and Reconceptions of an Extreme Atmospheric Hazard

John A. Knox, Brendan Mazanec, Emily Sullivan,
Spencer Hall and Jared A. Rackley

Additional information is available at the end of the chapter

http://dx.doi.org/10.5772/64019

Abstract

Superstorm Sandy was the second-costliest hurricane in U.S. history, causing catastrophic flooding and prolonged power outages in New Jersey and New York. The public's response to this extreme event on the social network Twitter is examined using statistical analysis and manual inspection of 185,000 "tweets" relating to Sandy. Sentiment analysis of tweets from Manhattan Island reveals a statistically significant trend toward negative perceptions, especially on the southern half of the island, as Sandy made landfall. Inspection of all tweets uncovered scientific misconceptions regarding hurricanes, and a surprising and disquieting anthropomorphic *reconception* of Sandy. This reconception, divorced from factual information about the storm, dominated the "Twittersphere" compared to official scientific information. The implications of such reconceptions for social media communication during future extreme events, and the utility of the methodology employed for analysis of other events, are discussed.

Keywords: hurricane, social media, scientific misconceptions, Superstorm Sandy, Twitter

1. Introduction

At the end of October 2012, Superstorm Sandy—a hurricane with the size of an extratropical cyclone—battered the mid-Atlantic coast of the United States. This chapter examines the response of the public on the social media venue Twitter [1] to the approach, landfall, and immediate aftermath of the storm.

Twitter is one of the most popular social networks in the world, with approximately 180 million users at the time of Sandy and over 300 million users at the end of 2015 [2]. A novel feature of Twitter is the restriction of posts to 140 characters or fewer. This restriction encourages brevity and contractions among users, and limits in-depth discussion, unlike other social networks such as Facebook.

The social media reaction to Sandy has been examined from a variety of angles. Edwards et al. analyzed millions of geolocated "tweets" (i.e., posts on Twitter) and focused on the utility of Twitter to meet needs often provided heretofore by first responders and relief agencies [3]. Similarly, Chatfield et al., in a conference presentation, studied the ability of Twitter users to convey time-critical information during this disaster [4]. Lachlan et al. noted, however, that Twitter communications during the storm were used more for emotional release than for dissemination of information, and that messages from official organizations were largely absent [5]. In addition, a large automated effort to examine Twitter messages during Sandy is underway at the National Center for Atmospheric Research [6].

Our research presented in this chapter attempts to combine the best aspects of statistical analysis of a large dataset with qualitative insights gained by manual, not automated, examination of individual tweets. We have accomplished the latter via the creation of original software which makes visual inspection of Twitter messages easy and efficient. Our initial objective was to characterize misconceptions and their propagation in Twitter posts; however, our research revealed not only misconceptions but an intriguing, and disquieting, *reconception* of Sandy that threatened to drown out factual messages regarding the storm.

Below, we provide an overview of Superstorm Sandy, discuss in detail the two methodologies used to study the dataset of tweets, examine the results from both methodologies, and provide conclusions based on our quantitative and qualitative results.

2. Physical science overview

This overview of Sandy relies on the authoritative post-storm analysis of [7]. Sandy began as a tropical wave off the west coast of Africa on 11 October 2012. After traversing the tropical Atlantic and moving westward into the Caribbean without much growth, it reached tropical storm intensity (34 kt or 39 mph or 17 ms^{-1}) on 22 October south of Jamaica (**Figure 1**). At this point the storm received the name "Sandy," from the list created by the World Meteorological Organization for Atlantic hurricanes which alternates between male and female names familiar to the cultures that border the tropical Atlantic. "Sandy" was selected a female name, given to the 18th named storm of the 2012 Atlantic hurricane season, following "Rafael."

Two days later, after executing a loop south of Jamaica, Sandy reached Category 1 hurricane strength (64 kt or 74 mph or 33 ms^{-1}) on the Saffir-Simpson Hurricane Intensity Scale at 1200 UTC 24 October just off the southeast coast of Jamaica. It then crossed Jamaica, moving northward, and became a major hurricane (100 kt or 115 mph or 51 ms^{-1}), i.e. a Category 3 on the Saffir-Simpson scale, just prior to landfall in Cuba. Jamaica, Cuba, Haiti, and the Dominican Republic were all impacted by Sandy, with 69 deaths and hundreds of thousands of homes destroyed, particularly in Cuba. Sandy expanded greatly in size after crossing Cuba, with the radius of tropical-storm-force winds doubling by the time it passed the Bahamas. However, its winds dropped in intensity, below the threshold for a hurricane.

Figure 1. The track of Sandy from inception in the Caribbean Sea until its demise after extratropical transition (ET) over the northeastern United States. Saffir-Simpson categories are indicated by number and shading. Image adapted from https://coast.noaa.gov/hurricanes/?redirect=301ocm.

As it moved northward in the western Atlantic, Sandy regained strength and became a menace to the Atlantic coast of the United States. Due to interactions with an approaching trough over North America and the influence of warm Gulf Stream waters underneath it, Sandy intensified as it moved north and reached a secondary maximum of 85 kt (98 mph or 44 ms^{-1}) at 1200 UTC on 29 October about 220 nautical miles (405 km) southeast of Atlantic City, New Jersey. Under the influence of a strong high-pressure system to its north, Sandy executed a highly unusual left turn that brought the storm inland over New Jersey instead of heading eastward out to sea. At 2330 UTC on 29 October, Sandy made landfall just northeast of Atlantic City with an estimated sustained wind of 70 kt (81 mph or 36 ms^{-1}) and a minimum central pressure of 945 mb. It was one of the largest and most intense hurricanes ever observed in the mid-Atlantic, particularly for so late in hurricane season. Due to the increasingly extratropical nature of the storm as it approached the coast, hurricane experts reclassified Sandy as a "post-tropical" storm shortly before landfall. However, in terms of hazards such as wind, rain, and storm surge, Sandy was virtually indistinguishable from a hurricane. In this paper we adopt the descriptor "Superstorm" for Sandy in order to reflect its dual nature.

Figure 2. The Battery Park underpass at the southern tip of Manhattan Island immediately after Sandy (top), and what it normally looks like (bottom). Photo courtesy K.C.Wilsey/FEMA.

Sandy's impacts in the United States were unusually severe and widespread, owing to its nearly 1000-mile swath of gale-force winds near the time of landfall. It was the deadliest tropical cyclone to hit outside of the southern U.S. in 40 years, with 72 U.S. deaths directly attributed to Sandy. A majority of these deaths (41 out of 72, or 57%) were due to storm surge along the Atlantic coast, but 20 of the deaths (28%) were due to falling trees. In addition, 87 indirect deaths were caused by Sandy, mostly related to loss of power during cold weather. At least 650,000 homes were damaged or destroyed in the U.S., approximately half in the state of New York near or along the coastline. So many homes perished because of the exceptional storm surge, aided by a full-moon tide, which reached a record 9.40 feet (2.87 m) above normal at the Battery on the southern tip of Manhattan Island (**Figure 2**). Similar high tides occurred just to the right of Sandy's eye at landfall; before it failed, the Sandy Hook tide gauge recorded a surge of 8.57 feet (2.61 m) above normal. Throughout the region, from New Jersey to Connecticut, barrier islands were inundated and in some cases breached; coastal areas were flooded to depths of several feet.

Superstorm Sandy caused approximately $50 billion in damage and was the second-costliest hurricane in U.S. history, topped only by Hurricane Katrina in 2005. About 8.5 million people lost electrical power during the storm, most of them in the hard-hit regions of New Jersey and New York; some customers were without power for months. It was the worst disaster in the history of the New York subway system. The flooding and power outages due to Sandy also closed the New York Stock Exchange for 2 days, the NYSE's longest closure in 124 years. It is in the context of this extreme event that we now turn to the public's response on Twitter.

3. Data and methodology

As a reminder, data for this study are Twitter posts, which are uncensored public utterances on a social media platform. Readers are advised of more-than-occasional strong language that is inevitably included in this narrative.

According to Pew Research, there were over 20 million tweets about Superstorm Sandy from October 27, 2012 through November 1, 2012 [8]. An analysis of that volume of data was beyond the scope of this project. Instead, we chose to isolate subsets of this immense trove of tweets and eventually created our own software to examine, both qualitatively and quantitatively, a sizable subset of the full trove. The intent was to create an intermediate level of breadth of tweets and depth of analysis of the tweets, rather than either to crunch statistics on a huge dataset or else to scrutinize in fine detail a small number of tweets. Our approach merges both statistical analysis with informed qualitative impressions based on the personal reading of thousands of tweets, made efficient via our software.

To this end, a third-party Twitter export service, GNIP Company, was used to acquire the necessary Twitter data needed for this study. The service used the two keywords "sandy" and "superstorm" to sift through all of the tweets posted between October 25, 2012, and November 2, 2012, tag the tweets that contained one or both of those keywords, and export the tagged tweets to two JavaScript Object Notation (JSON) files. A JSON file is a representation of JavaScript Objects in text form. These files are referred to below as "the full dataset" or "the complete dataset," and contain approximately 185,000 individual tweets. To our knowledge this wealth of data makes our research one of the more comprehensive analyses to date of social media during the Sandy event.

Two methodologies were pursued in the course of our research, and are discussed in detail below.

3.1. Sentiment analysis methodology

3.1.1. Data

Initial analysis of the dataset was begun using the open source software OpinionFinder [9], which can identify subjectivity and positive or negative sentiment in phrases. This software is used widely in multiple disciplines [10]. Tweets are classified on a numeric scale with positive tweets set to greater than zero, negative tweets set to less than zero and neutral tweets set to zero. Problems inherent in the OpinionFinder classifying system are described in Ref. [11]. This study assumes that OpinionFinder correctly identifies tweets as positive, negative, or neutral, but also notes the tendency of OpinionFinder to over-classify tweets as neutral. Fortunately, a neutral zero does not skew the data. Numerous other studies have used OpinionFinder analysis of tweets to conduct research, including some work on Superstorm Sandy [12].

A small subset of the full dataset, tweets on Manhattan Island, was examined. The point of this component of the research was to track the evolution of sentiment over time; therefore, classified tweets were divided into equal time intervals. The first analysis uses 17 time intervals of 12 hours each, while the second compares a before-event interval of to a during/after event period. Twelve-hour intervals were chosen for simplicity while maintaining high enough time resolution to denote change. The two 12-hour periods from 1200 UTC 29 October to 1200 UTC 30 October also encompass an appropriate time period of the direct impacts from Sandy on Manhattan. The time of 1200 UTC 29 October was chosen as the mark between pre-event and event based on Manhattan weather observations.

Classified tweets for each time period were then aggregated using ArcMap to census tracts in Manhattan, with each tract taking the sum of the classified tweets (i.e., tracts that have a higher proportion of positive tweets to negative tweets obtain a more positive rank). Aggregating the data points to census tracts made it more manageable and help to smooth out pockets of many tweets versus areas with fewer tweets. Census tracts also help to distribute the data by population, because tracts are roughly similar in population count. The census tracts in Manhattan are small enough in size that it can be assumed residents within the area experienced similar effects from Sandy.

Bias is inherent in using data such as tweets. Here, the database is obviously biased towards those with access to the internet via desktop (and especially access to smartphones, because power outages occurred), and is presumably also biased toward a younger demographic that utilizes Twitter. For this study, we assume that the opinions reflected in the tweets adequately reflect the opinion of others in the area that did not have access to Twitter.

3.1.2. Statistical analysis

The first step in analyzing the evolution of the opinions over time was to determine if there is a statistically significant change. To accomplish this, an analysis of variance (ANOVA) model was run using the RStudio statistical IDE [13]. ANOVA can effectively determine if one or more of the time intervals has a statistically different positive/negative opinion. In our work, the null hypothesis was that no change in opinion occurred over the 9 days.

To further assess an evolution in opinion, a second test was used to determine if there was a difference in opinion between two time periods: before Sandy's impacts were felt in Manhattan and after the effects began to be felt. In order to determine which test is appropriate for measuring the difference between these time pairs, that data were tested for normality using the Kolmogorov-Smirnov (K-S) Goodness-of-Fit test. If the data were determined to be non-normal by the K-S test, then the Wilcoxon Matched Pairs Signed-Ranks test was then used to determine a difference. (The Wilcoxon Matched Pairs Signed-Ranks test is appropriate for determining whether there are significant differences in a pair of non-parametric data such as this.) Again, the null hypothesis was that there was no change in opinion before and after Sandy's impacts. Finally, sums of the classified opinions for each of the 12 time intervals were plotted using R's barplot command. This allowed for visualization of the data and further determines directionality of the alternate hypothesis.

3.2. TweetReviewer methodology

3.2.1. Data

After our initial OpinionFinder effort, we decided to explore the full dataset in more breadth and depth. The JSON files were then imported to a MySQL database using an import program written in PHP to accomplish this more in-depth analysis.

3.2.2. Quantitative and qualitative analysis

Once the data were in the MySQL database, a tool was needed to aid in reviewing the tweets, sorting them as relevant (pertaining to Superstorm Sandy in some manner) or irrelevant (not referring to Superstorm Sandy) to the project, and bookmarking the tweets of interest. The program TweetReviewer was built by the second author on the .Net 4.0 Framework and written in C# for this purpose (**Figure 3**).

A set of filter words were created and plugged into the program to help determine the relevance of the numerous tweets. The filter words are listed in the Appendix. These filters were used to aid the researchers as they went through the tweet dataset by hand to determine the relevance of the tweets the filters did not tag. The tweets deemed relevant either through the filter tagging process or by hand, were sorted together. The non-relevant tweets were also marked accordingly and separated from the relevant tweets. However, none of the non-relevant tweets were deleted. The entire database of tweets was archived and retained.

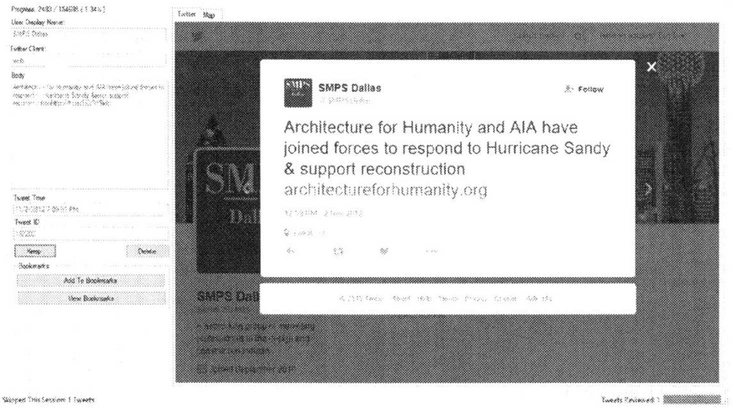

Figure 3. TweetReviewer software graphical user interface.

Once the relevant and non-relevant tweets were separated, new keywords were searched using MySQL's search capabilities to identify the number of times that particular keyword was used. This process helped provide a clearer picture of the perception of Superstorm Sandy through the lens of Twitter.

The posts have been analyzed spatially as well as temporally, with word counts and word clouds as a function of day and time scrutinized as quantitative measures of public responses. We have also examined tweets individually by the thousands using the database, which has permitted the authors to develop qualitative insights into the public response that would be difficult or impossible without simple visualization and bookmarking tools.

4. Results

4.1. Sentiment analysis

Analysis using ANOVA allowed for a rejection of the null hypothesis that no change in opinion occurred over the 9 days examined. ANOVA showed that there was a significant difference between the opinions of the different 12-hour time intervals. As seen in **Table 1**, based on an F-value of 3.971 calculated using [14] and a p-value of 0.0464, a difference was found between the time intervals that are significant at the 95% confidence level.

The second set of tests included the Kolmogorov-Smirnov (K-S) Goodness-of-Fit test, and Wilcoxon Matched Pairs Signed-Ranks test. Results from the K-S tests, shown in **Table 2**, demonstrate that the data are not normally distributed. p-Values for both "before" and "after" data are vanishingly small (<< 0.001), indicating that the null hypothesis can be rejected and a non-parametric test statistic would be more appropriate, such as the Wilcoxon Matched Pairs Signed-Ranks test.

4.2. TweetReviewer analysis

The results uncovered by sentiment analysis failed to capture other insights that the researchers felt were important but were outside the scope of OpinionFinder. This led to the creation and use of TweetReviewer software to allow us to examine, both quantitatively and qualitatively, other aspects of the full database.

While analyzing the full database, it was determined there were two major common categories of interest with regard to the public's perceptions of Superstorm Sandy: scientific misconceptions, and an anthropomorphic reconception of Sandy that, in the parlance, "went viral."

4.2.1. Misconceptions

There were many misconceptions regarding Hurricane Sandy and hurricanes in general in the dataset. One quotidian misconception was the common misspelling of the word "hurricane." In fact, the word "hurricane" was misspelled so often that the list of filtered words had to be adapted to accommodate the many misspellings (e.g., see #51 in Appendix). More substantively, there was a general lack of understanding of what defines a hurricane. These misconceptions can be separated into four categories: *strength, category, size,* and *duration.*

The strength of a hurricane is most commonly associated with the Saffir-Simpson Hurricane Wind Scale categories. Overall, the Twitter users had a fairly decent grasp of the numerical scale as far as the numbers were concerned, but the understanding of what those numbers stood for was almost entirely absent. Throughout the dataset (**Table 4**) were tweets, many from Florida, that downplayed Hurricane Sandy claiming there was little to worry about because it was "only cat 1." Instead of acknowledging the force of the hurricane, users discriminated against the lower numbers even though the Saffir-Simpson Scale is based on wind only. It does not take into account the damage from other impacts like the storm surge and rainfall, and neither did the Twitter users. A few tweets mentioned Hurricane Sandy as having a storm surge more often associated with a Category 3 or Category 4 hurricane, such as this (tweet #76357 in the full database):

> "Irene was a category 3. Sandy is a 1 but with the storm surge it's supposed to act as a 4 at most."

	Category 1	Aftermath	Apocalypse	Worse	Power
25 Oct. 2012	2	0	1	6	10
26 Oct. 2012	13	1	4	28	28
27 Oct. 2012	5	1	4	21	55
28 Oct. 2012	17	7	343	61	182
29 Oct. 2012	26	8	1136	213	1326
30 Oct. 2012	14	212	398	220	1748
Total	**77**	**229**	**1886**	**549**	**3349**

Table 4. Tweet word counts related to hurricane strength.

	Tornado	Touchdown	Landfall	How Long
25 Oct. 2012	13	0	10	0
26 Oct. 2012	21	0	11	1
27 Oct. 2012	33	1	10	1
28 Oct. 2012	126	7	22	2
29 Oct. 2012	213	4	142	13
30 Oct. 2012	170	2	43	10
Total	**576**	**14**	**238**	**27**

Table 5. Tweet word counts related to hurricane size.

	Over/over yet	Finish	Finally	Bring it	Passed/past
25 Oct. 2012	28	1	0	2	2
26 Oct. 2012	69	2	3	13	3
27 Oct. 2012	76	2	6	35	6
28 Oct. 2012	244	17	17	112	22
29 Oct. 2012	784	38	60	191	60
30 Oct. 2012	785	30	67	23	82
Total	**1986**	**90**	**153**	**376**	**175**

Table 6. Tweet word counts related to hurricane duration.

The size and duration of Hurricane Sandy was also frequently misunderstood (**Tables 5** and **6**). Very few tweets implied that the users fully grasped the sheer size of any hurricane, let alone the immense size of Sandy. They appeared to assume that a hurricane was a small storm that would be bad, much like a supercell or "derecho-sandy thing" (tweet #4885) that would "touchdown" or "touch land" (tweet #100800), wreak havoc, close schools, and leave, all within a matter of hours. There were several tweets that asked if the storm was "over yet?" Hardly any of the Twitter users seemed to grasp the fact that a hurricane is actually a huge storm spanning hundreds of miles which can last for days. Some tweets even called Hurricane Sandy a tornado (tweets #38455 and #49032) or, as one CEO put it, a "tornadocaine" (tweet #5267).

Among many visual misconceptions propagated during Superstorm Sandy, one of the most prominent and widely shared was a Photoshopped image of a supercell thunderstorm over the Statue of Liberty in New York City (e.g., tweet #60774). The supercell thunderstorm was indicated to be Sandy—another sign of confusion regarding the vast differences in size, strength, and duration between hurricanes and tornadic thunderstorms. This misconception became so popular on social media that a story about it appeared in the Philadelphia media [16].

4.2.2. *The reconception of Sandy*

As Superstorm Sandy approached the mid-Atlantic coast, a number of Twitter users began to do something peculiar, at least from the perspective of scientists or emergency managers. Instead of relaying factual information regarding the storm, accounts pretending to be the personification of Hurricane Sandy started appearing, as venues for posting jokes about the hurricane from a first-person perspective. With the help of these accounts, the Twitter community took the idea of personifying Sandy and expanded upon it. Without any visible evidence of premeditation regarding the nature of this anthropomorphic Sandy, the Twitter community banded together and simply accepted a persona of their creation without question. They also hurled insults at the hurricane based on its fabricated persona and its perceived status as a female.

Sandy was suddenly no longer an impersonal, inanimate hurricane; she was an "independent sassy black hurricane who don't need no man" (tweets #13365 and 29 subsequent tweets), who grew up in the ghetto, went to school with (Hurricane) Irene, cursed up a storm, voted Democratic, knew how to work the pole, had an avid and colorful sex life, and was the brunt of many cruel taunts related specifically to female anatomy. **Table 7** provides a sampling of word counts related to this persona in the days leading up to landfall.

	Bitch	Whore	Twerk	Pole	Shark	Black	Nigga
25 Oct. 2012	8	0	0	0	2	0	0
26 Oct. 2012	35	2	0	1	0	1	7
27 Oct. 2012	61	3	4	0	0	8	3
28 Oct. 2012	253	17	25	2	6	22	29
29 Oct. 2012	614	41	42	16	32	96	68
30 Oct. 2012	382	31	14	21	74	168	39
Total	**1353**	**94**	**85**	**40**	**114**	**295**	**146**

Table 7. Tweet word counts related to the invented persona of Sandy.

The various Twitter accounts created to portray this user-crafted persona encouraged its incredibly fast proliferation and popularity across the website. They posted tweets which consisted mainly of rather dirty jokes, inappropriate suggestions, racial slurs, female-specific insults, and Republican-specific insults. One of the aforementioned tweets talked about the Sandy persona tossing a trailer at a woman in a minivan simply because she had a bumper sticker of a Republican presidential candidate on her car (tweet #21587).

It is possible that the timing of the hurricane's landfall, within a week of the 2012 presidential elections when tensions between the Democratic and Republican parties were already high, amplified the politically slanted comments. But the outright cruelty and twisted content went beyond simple political reasons. To make matters worse, these offensive tweets were shared hundreds of times by users who were apparently from a variety of ages, races, and political beliefs.

The tweets referring to the Hurricane Sandy persona consistently referred to her as a "bitch" and joked about her coming to "blow" the entire east coast and make everyone "wet" or simply told her to "fuck off." This was such an exceedingly common theme that the words "bitch," "blow," "wet," and "fuck" became keywords for both the filters during the process of going through the tweets as well as for searching their word counts. Of these four words, "wet" was used the least common, only garnering a word count of 544 times used. The word "blow" was used 1129 times, "fuck" was used 3655 times, and "bitch" led all epithets in the full database with a word count of 4335 times used with regard to Hurricane Sandy.

These word counts are, of course, small compared to the total number of words used in the full database. To give a sense of how the persona of Sandy dominated the "Twittersphere," **Table 8** presents the fractional representation of scientific/hurricane-related terms in the full database versus the top three persona terms. As shown in the table, the persona was many times more popular on Twitter than were factual reports about Sandy.

NWS	MPH	Category/cat	Landfall	Windspeed
0.022	0.133	0.387	0.079	0.0001

Table 8. Ratio of word count of scientific terms to the word counts of the three most common Sandy-persona terms.

5. Conclusions

Our results shed light on public perceptions, misconceptions, and reconceptions of an extreme atmospheric hazard. Superstorm Sandy was a virtually unprecedented event along the Atlantic coast of the United States, in terms of intensity, size, and path. What did the public make of this event on Twitter?

From our analysis of Manhattan Island data, it is plausible that residents underrated the storm's ferocity until the last 24 hours before landfall. The largest sentiment swings occurred, quite naturally, in and near the regions most affected by the storm: the coastline and the south Manhattan neighborhoods blacked out due to a power plant failure.

Our more fine-grained analysis using the TweetReviewer software revealed additional aspects of the public's reaction to Sandy. Profound confusion exists regarding the size, strength, and duration of hurricanes. The public seems to confuse hurricanes with tornadoes; this confusion during Sandy extended to the Photoshopped image of a supercell thunderstorm over the Statue of Liberty. Expectations that Sandy would be as brief and intense as a tornado were not met; in particular, the unusually large extent of Sandy (the largest storm in terms of diameter of gale-force winds since records began in 1988; see [7]) was not well understood by those on

Twitter. The more subtle point that Sandy could be "only a Category 1" and still do extensive damage due to storm surge was also not grasped.

What the "Twittersphere" did seem to eagerly grasp was a user-generated anthropomorphic "Sandy" who dealt out death and destruction like a villain in a superhero comic or movie. This personified "Sandy" was then made the object of race-based and gender-based slurs that were widely perceived to be amusing, rather than offensive if they had been made in public. This created an increased level of noise that tended to drown out the true signal: reports of important developments in Hurricane Sandy's category changes, the watches, advisories, and warnings, and tweets with legitimate scientific information that could have better informed the public. Our results thus align more with those of [5] than with the more positive, life-saving impacts of Twitter found by other researchers. Our analysis of geo-located tweets may bias our results somewhat in this respect, however (R. Morss, pers. comm., January 2016).

When much of New York City lost electricity, the theme of attacking the hurricane based on its manufactured persona quickly fell by the wayside in favor of complaining about the power loss and realizing Sandy was in fact a hurricane and not a fictitious creation. The persona continued to play a substantial role in the dataset, but the newest point of interest was the loss of power. In fact, the word "power" was found 8649 times in our database, making it more popular than "bitch."

We conclude, with some surprise, that until the hurricane interfered directly with people's personal lives, Twitter users seemed content with obsessing over the invented persona of Hurricane Sandy. Rather than bemoaning this flight from reality, however, we encourage a more proactive response among emergency management personnel, meteorologists, and others who communicate directly with the public. Perhaps it is possible that this behavior can be utilized for the benefit of the public. If the public could quickly create and propagate a persona for a hurricane, it stands to reason that official outlets could do the same. They could create a Twitter account for a new hurricane, create their own persona for it, and use this to disseminate the important relevant information to the public in a format that is more easily digestible than more esoteric scientific criteria.

For example, a large swath of the U.S. population is familiar with superheroes on some level, whether DC, Marvel (or for even younger audiences, Pokémon). What if scientists began using popular superheroes to help describe the strength of a storm? As outlandish as it sounds, this is something the general public would grasp fairly easily, simply because they are already familiar with the characters. For instance, intense heat could be described as being on par with an attack by Marvel's Human Torch, a member of the Fantastic 4 superhero group. A powerful electrical storm could be compared to Marvel superhero Thor wielding Mjolnir in battle. The incredible storm surge of a hurricane such as Sandy could be compared to the DC supervillain Ocean Master or New Wave in combat. Similar translations of specific hazards into personae would be possible in Pokémon, reaching even younger audiences.

These examples may come across as childish, but the public would easily grasp a general idea of the intensity of the heat, the electrical storm, and the storm surge from them. More importantly, children would easily comprehend these examples. A parent can disregard weather alerts, but not if their children continually bother them. If this method of using pop-culture references to explain something as confusing as weather can bridge the communication gap to the next generation, then they could learn to listen to and obey weather alerts. This "if you can't beat 'em, join 'em" approach to anthropomorphizing atmospheric hazards could capture some of the social media energy that might otherwise propel a completely non-factual (and offensive) personification to prominence, as occurred with Sandy.

Finally, we advocate the use and/or development of software such as our TweetReviewer as a means for visually inspecting thousands of tweets easily and efficiently. As our research indicates, actually reading the tweets provides insights that are unlikely to be gained by mere statistical crunching on datasets. The filtering capabilities of TweetReviewer enable the user to focus on relevant tweets and screen out non-relevant tweets, significantly accelerating the process and permitting human analysis of relatively large datasets.

Acknowledgements

The first author gratefully acknowledges the support of an M.C. Michael Award from the University of Georgia, which funded much of the research presented here.

Appendix

Filter words used in our TweetReviewer software to narrow the database to the most relevant posts are:

1. "red cross" or "redcross"
2. "#sandy" if either "springs" or "beach" is not present.
3. "#hurricanesandy" or "#hurricainesandy"
4. "hurricane sandy"
5. "jersey shore"
6. "#njsandy"
7. "hurricane death megatron"
8. "sandy aftermathpocalypse"
9. "halloweenpocalypse"
10. "#postsandy"
11. "#aftersandy"
12. "#fucksandy"
13. "#fuckyousandy"
14. "#fusandy"

15. "#fyousandy"
16. "fuck sandy"
17. "fuck you sandy"
18. "#ctsandy"
19. "#bptsandy"
20. "#nhsandy"
21. "#dcsandy"
22. "#wvsandy"
23. "#nysandy"
24. "#ny1sandy"
25. "#nycsandy"
26. "#masandy"
27. "#vasandy"
28. "#mdsandy"
29. "#vtsandy"
30. "#tosandy"
31. "#6abcsandy"
32. "#lifeaftersandy"
33. "storm sandy"
34. If the tweet contains both "Sandy" and "Katrina"
35. If the tweet contains both "Sandy" and "Irene"
36. "#thankshurricanesandy"
37. "#thankssandy"
38. "#thankyousandy"
39. "#damnyousandy"
40. "#damnitsandy"
41. "post-sandy"
42. "post sandy"
43. "#frankenstorm"
44. "superstorm"

45. If the tweet contains both "Obama" and "Fema"
46. "fema"
47. "#damnsandy"
48. "#darnsandy"
49. "#ihatesandy"
50. If the tweet contains both "prayers" and "affected by sandy"
51. "@hurricannesandy"
52. "@ahurricanesandy"
53. "@sandydahurricane"
54. "@sandyshurricane"
55. "#survivingsandy" or "#survingsandy"
56. "#stormsandy"
57. "sandy storm"
58. "Frankenstorm"
59. "hurricane"
60. "bitch sandy"
61. "#pray"
62. "#hurricane"
63. If the entire tweet is "superstorm sandy xuo"

Author details

John A. Knox[1*], Brendan Mazanec[1], Emily Sullivan[1], Spencer Hall[1] and Jared A. Rackley[2]

*Address all correspondence to: johnknox@uga.edu

1 University of Georgia, Athens, GA, USA

2 Oak Ridge National Laboratory, Oak Ridge, TN, USA

References

[1] Twitter. [Internet]. 2016. Available from: https://www.twitter.com [Accessed: 2016-04-03]

[2] Statista, The Statistics Portal. [Internet]. 2016. Available from: http://www.statista.com/statistics/282087/number-of-monthly-active-twitter-users/ [Accessed: 2016-04-03]

[3] Edwards JF, Mohanty S, Fitzpatrick P, Martin D. Assessment of social media usage during Hurricane Sandy and the development of a Twitter-based, web application for real-time communication of storm-related information. 96th American Meteorological Society Annual Meeting. New Orleans, LA, 11 January 2016. https://www.hpc.msstate.edu/publications/docs/2016/01/14285social_media_tool_storm_info_ams.pdf [Accessed 2016-04-03]

[4] Chatfield AT, Scholl HJ, Brajawidagda U. #Sandy Tweets: Citizens' co-production of time-critical information during an unfolding catastrophe. Proceedings of the 47th Hawaii International Conference on System Science, IEEE Computer Society, Waikoloa, Hawaii. 2014. pp. 1947-1957.

[5] Lachlan KA, Spence PR, Lin X, Del Greco M. Screaming into the wind: Twitter use during Hurricane Sandy. *Communication Studies*. 2014; 65 (5): 500-518.

[6] Demuth JL, Morss RE, Palen L, Anderson K, Anderson J, Kogan M, Stowe K, Bica M. Understanding dynamic communication, risk perception, and decisions during Hurricane Sandy through analysis of Twitter data, Society for Risk Analysis annual meeting, Arlington, VA, 9 December 2015.

[7] Blake ES, Kimberlain TB, Berg RJ, Cangialosi JP, Beven III JL. Tropical cyclone report: Hurricane Sandy (AL182012) 22-29 October 2012. [Internet]. 2013. Available from: http://www.nhc.noaa.gov/data/tcr/AL182012_Sandy.pdf [Accessed: 2014-04-03]

[8] Pew Research Center. [Internet]. 2013. Available from: http://www.pewresearch.org/fact-tank/2013/10/28/twitter-served-as-a-lifeline-of-information-during-hurricane-sandy/ [Accessed: 2016-04-03]

[9] OpinionFinder. 2.x Release Page [Internet]. 2011. Available from: http://mpqa.cs.pitt.edu/opinionfinder/opinionfinder_2/ [Accessed: 2016-04-03]

[10] Bollen J, Mao H, Zeng X-J. Twitter mood predicts the stock market. *Journal of Computational Science*. 2010; 2 (1): 1–8. doi:10.1016/j.jocs.2010.12.007

[11] Wilson T, Hoffman P, Somasundaran S, Kessler J, Wiebe J, Choi Y, Cardie C, Riloff E, Patwardhan S. OpinionFinder: A system for subjectivity analysis. Proceedings of Human Language Technologies Conference/Conference on Empirical Methods in Natural Language Processing (HLT/EMNLP 2005), Vancouver, Canada, 2005. pp. 35-36.

[12] George Mason University. The GeoSocial Gauge [Internet]. 2016. Available from: http://geosocial.gmu.edu/projects/superstorm-sandy/ [Accessed: 2016-04-03]

[13] R Core Team. R: A language and environment for statistical computing. R Foundation for Statistical Computing, Vienna, Austria, 2015. URL https://www.R-project.org/.

[14] Free Statistics Calculators. Version 4.0 [Internet]. 2016. Available from: http://www.danielsoper.com/statcalc3/calc.aspx?id=4 [Accessed: 2016-04-03]

[15] New York Magazine. From The Editors [Internet]. 2012. Available from: http://nymag.com/nymag/letters/hurricane-sandy-editors-letter-2012-11/ [Accessed: 2016-04-03]

[16] Mathis J. That Awesome Hurricane Sandy Photo of the Statue of Liberty? Fake [Internet]. 2012. Available from: http://www.phillymag.com/news/2012/10/29/awesome-hurricane-sandy-photo-statue-liberty-fake/ [Accessed: 2016-04-03]

Chapter 3

Communication and Preparedness Issues on Various Scales from Extreme Atmospheric Hazards

Robert M. Schwartz

Additional information is available at the end of the chapter

http://dx.doi.org/10.5772/63447

Abstract

Most federally declared disasters are from atmospheric hazards. These could be from floods, tropical cyclones, tornadoes, or winter storms. Some of these hazards are events with relatively short warning times such as tornadoes or have sufficient warnings from tropical cyclones. This research examined communication of weather information and personal preparedness following the Florida landfall from Tropical Storm Debby in 2012. Another case study examined emergency management issues such as preparedness and response after the 2011 tornado in Tuscaloosa, Alabama. The concentration was on emergency management agencies at the County, University, State, and Federal levels. A sample of elderly residents of Pinellas and Pasco Counties in Florida completed a self-administered survey to examine various means of receiving weather information along with hurricane preparedness actions. In-depth interviews were conducted with representatives of various agencies on different scales in regard to preparedness and response following the Tuscaloosa tornado. The elderly used television as the primary means of receiving weather information, thus stressing the importance of utilizing both traditional and newer forms of communications to reach all citizens. One of the major issues on all levels following the Tuscaloosa tornado is related to communications such as resource allocations and response actions.

Keywords: disaster communications, tornado and hurricane preparedness, personal and community emergency management preparedness, methods of receiving weather information, emergency management response

1. Introduction

Most of the federally major declared disasters are due to atmospheric hazards. Of the 3477 total declarations, from 1953 to 2015, 2449 or 71% are weather related [1]. There were 293 major disaster declarations from 2011 to 2015. Only 18 of these major declarations were non-weather related such as earthquakes [2]. Meteorological disasters could be from events such as but not limited to floods, tropical cyclones, tornadoes, or winter storms. Some of these hazards are quick fuse events with relatively short warning times such as tornadoes or have sufficient warning such as a tropical cyclone. This research examined two different atmospheric hazards that were declared major disasters. The first studied means of communicating weather information and personal preparedness of elderly citizens following Tropical Storm Debby in 2012 that made a landfall in Florida. An additional case study examined emergency management issues such as preparedness, response, recovery, and mitigation after the tornado in Tuscaloosa, Alabama, in 2011. The research covered emergency management organizations at the county, university, state, and federal levels.

2. Background

2.1. Tropical cyclones

Some of the most destructive storms are tropical cyclones. Based on the geographic location, tropical cyclones are known as typhoons over the western Pacific Ocean, cyclones over the Indian Ocean, and hurricanes over the Atlantic and eastern Pacific Oceans. Hurricane season in the North Atlantic basin (which impacts the United States) starts 1 June and ends 30 November, with the peak being around mid-September. The North Atlantic basin includes the Gulf of Mexico, Caribbean Sea, and Atlantic Ocean [3,4].

Tropical cyclones are classified by organization of thunderstorm clusters, circulation patterns, and wind speeds. First there is a tropical wave which is an unorganized cluster of thunderstorms with a weak surface circulation. An organized cluster of thunderstorms is generally 150–350 miles (250–600 km) in diameter with no closed circulation and maintains an identity for 24 hours is a tropical disturbance. A tropical depression has an identifiable pressure drop, closed circulation, and wind speed less than 39 mph (34 kts); the system is then assigned a number by the National Hurricane Center. If the storm continues to organize with wind speeds greater than or equal to 39 mph (34 kts) to less than 74 mph (64 kts), it is classified as a tropical storm and given a name. Continued development of the system becomes a hurricane with wind speeds exceeding or equal to 74 mph (64 kts) [3,4].

There are five major environmental factors which determine successful tropical cyclone development. These are sea surface temperature, surface layer of warm water, weak vertical wind shear, sufficient moisture in the middle troposphere, and a location at least 5° north or south of the equator. Sea surface temperature must be greater than 80°F (26.5°C) as this supplies the heat and moisture released into the atmosphere. The layer of warm water is usually around

http://dx.doi.org/10.5772/63447

200 ft (60 m) to ensure there are enough warm water and energy as the ocean gets mixed in a process known as upwelling. This depth keeps the warmer water rising and cooler water in deeper regions that will not impede the energy source. Wind shear (winds from opposite directions or too high a speed) needs to be relatively weak to form the vortex. If speeds are too high, the vortex will be torn apart and moves downstream. Dry air in the middle troposphere can weaken the storm by reducing latent heat and increasing downdrafts through evaporative cooling; hence, a tropical cyclone requires enough moisture to ensure formation. Rotation is another important aspect of tropical cyclones and thunderstorm clusters. These clusters need to be 5° north or south of the equator so the Coriolis force is strong enough to help develop rotation; the Coriolis force is zero at the equator [4].

The Saffir-Simpson Scale classifies hurricanes based on wind speed on a scale of 1–5 with 5 the strongest. Before 2010, the scale predicted storm surge height and barometric pressure along with wind speeds. The wind scale was revised again in 2012 (**Table 1**) due to the rounding and conversions from mph to km/h; only category 4 and 5 storms were impacted with this update. Previous storms will not have their categories changed with the new scale. Expected damage generally increases with higher category levels. Category 3 and higher hurricanes are known as major hurricanes [4,5].

Category	Wind speed	Damage type
1	74-95 mph	Some
2	96-110 mph	Extensive
3	111-129 mph	Devastating
4	130-156 mph	Catastrophic
5	157 mph or higher	Catastrophic

Table 1. Saffir-Simpson scale [5].

Tropical cyclones are some of the most destructive storms. Some of the impacts are due to storm surge, heavy rain, inland flooding, high winds, and potential tornadoes. Storm surge refers to the rise sea level as the hurricane makes landfall. Onshore winds and the barometric effect (rise in sea level due to low pressure) cause storm surge. Other factors are the wave height, tides, and shoreline shape. Storm surge can cause extensive damage to the landscape and structures. If the tropical cyclone is slow moving or stalled, heavy rains can hit an area. Heavy rain can produce inland flooding and causes the most fatalities along with property destruction. High winds can cause damage, especially to structures not built to withstand higher tropical force winds. There can also be hurricane spawned tornadoes which are often highly concentrated in the right, front quadrant (when looking at the perspective of the hurricane approaching the shore). These tornadoes are usually in the EF0–EF2 range and can also cause property damage [4].

2.2. Tropical Storm Debby

During the 2012 hurricane season, Tropical Storm Debby originated in the south-central Gulf of Mexico after a surface low developed near the Yucatan peninsula and propagated eastward toward an area where the prevailing subtropical ridge had weakened. There was a northern edge of a tropical wave in the Caribbean Sea that merged with the disturbance near the Yucatan peninsula that eventually became Tropical Storm Debby on 22 June [6]. Hurricane Hunter aircraft determined circulation was well defined and the winds were tropical storm strength on 23 June (**Figure 1**). Over the next 24 hours, Tropical Storm Debby moved slowly north to northeastward without a well-defined trajectory forming a rain shield over the northeastern Gulf of Mexico on 24 June. On 25 June, Debby approached the Big Bend area of Florida (area where the Panhandle and Florida peninsula curve on the Gulf of Mexico) and made landfall near Steinhatchee, Florida, on 26 June. Peak winds were estimated at 63 mph (55 kts) and minimum surface pressure was 990 mb [6].

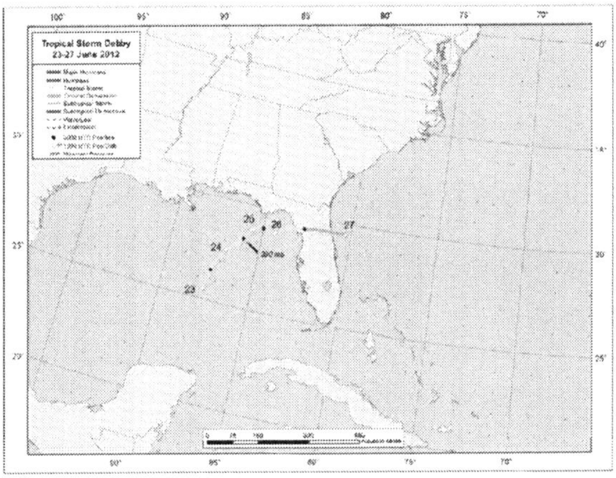

Figure 1. Track positions of Tropical Storm Debby [6].

Winds were not the major impact from Debby but the torrential rains caused major impacts along with inland flooding in various parts of the Florida peninsula. A local observer in Wakulla County measured 29 in. (731 mm) during the event and there were other reports greater than 20 in. (508 mm) in the same region (**Figure 2**). Wakulla County is northwest of the track line near the Great Bend with the light blue shading. There were several totals greater than 10 in. (254 mm) over western and northeastern Florida (see **Figure 2**, south of the purple-shaded track line) [6].

Besides inland flooding, storm surges from 2 to 4.5 ft (0.6–1.4 m) occurred from the Florida Panhandle to the southwestern coast of Florida. This resulted in inundation 1–3 ft (0.3–0.9 m) aboveground level with the highest surge reported between Apalachicola and Cedar Key [6] (see **Figure 2**) on and northwest of the purple-shaded track line) [6]. In addition to the heavy

Section 2

Part II: Atmospheric Dynamics and Modeling Aspects of Atmospheric Hazards

A Case Study of Atmospheric Dynamics and Thermodynamics in Derechos and the Societal Impacts

Kevin Law

Additional information is available at the end of the chapter

http://dx.doi.org/10.5772/63319

Abstract

The word "derecho" is used to differentiate a storm having straight-line winds as opposed to rotational, tornadic winds. Although the term "derecho" is relatively old, derechos were not readily recognized by the general public until recent outbreaks caused significant widespread damage and associated fatalities. Most notably, the 2012 Mid-Atlantic Derecho in the USA brought these types of storms to the public's attention as a variety of societal impacts including infrastructural damage, power outages, and fatalities occurred over an extensive area from outside of Chicago to Washington, DC The associated damage can be more widespread than tornadoes, and the number of fatalities is comparable to those found in medium-intensity tornadoes.

This study investigated the importance of the dynamics and thermodynamics in maintaining the intensity of derechos. Key meteorological parameters were measured over six stations where the 2012 Mid-Atlantic Derecho passed. Low- to mid-level wind shear, as well as the Convective Available Potential Energy (CAPE) and Most Unstable Convective Available Potential Energy (MUCAPE), was found to be significantly higher at the time of passage, which allowed the system to intensify and propagate downstream.

Keywords: derechos, dynamics, thermodynamics, convection, societal impacts

1. Introduction

North America is known for its many types of atmospheric hazards including tornadoes, thunderstorms, blizzards, and hurricanes. Physical geography plays a pivotal role as two mountain ranges, the Rockies in the west and the Appalachians in the east, are both oriented north-south. Since there is no physical barrier to prevent cold air masses to the north and warm air masses to the south from interacting with each other, severe and hazardous weather is commonly produced. The warm, moist air is easily lifted by the colder, drier air producing mid-latitude storms like no other place on Earth. In fact, North America generates almost 75% of all known tornadoes [1].

However, there is another type of atmospheric hazard that does not rotate violently but nevertheless can cause considerable damage and a variety of societal impacts—the *derecho*. The term "derecho" is not as commonly known as "tornado" but has been around for many years. Derecho was first used by a University of Iowa physics professor, Dr. Gustavus Hinrichs, in 1888 to describe a thunderstorm that produced strong straight-line winds as opposed to rotational winds [2]. Hinrichs believed another term was needed to differentiate violent straight-line storms from tornadoes; therefore, he used "derecho," which is Spanish for "straight ahead." However, the word "derecho" was virtually absent from the meteorological lexicon for almost 100 years until the 1980s when Robert Johns and William Hirt [3] re-introduced the term. Although, not as widely known as tornadoes by the general public, derechos are gaining recognition for their destructive capability [4]. Even though derechos can form anywhere around the world, they predominantly form in the USA. In fact, very few studies have even been published about derechos occurring outside North America [5, 6]. Therefore, the focus of this study will be on the USA and a particularly strong derecho in 2012.

2. Background

Derechos have been specifically defined as "families of downburst clusters" that originate from a mesoscale convective system while covering an area where the major axis is at least 400 km [3, 7, 8]. They are associated with causing straight-line, non-tornadic damage and occur most frequently during the summer months, especially east of the Rocky Mountains in the USA [3, 9]. In addition, areas of convection produce "bow echoes" [10] that propagate downstream and can create severe downbursts.

2.1. Types of derechos

Derechos are often divided into four different types: progressive, serial, hybrid, and low-dew point (dry). Regardless of the type, derechos often form from a mesoscale convective system where the individual thunderstorms often start to replicate as they propagate downstream.

Progressive derechos are often found along a stationary frontal boundary oriented west-east with mid-level winds flowing parallel (west-east) to the frontal boundary [3, 11]. With sufficient convection, a mesoscale convective system will develop along the boundary evidenced by a bow echo typically having a length of up to 250 miles in the beginning phases. As the system travels downstream, a sharp downdraft of cold air intensifies near the center of the bow echo and is pushed further ahead by the west-east mid-level wind flow. The size of the derecho often increases to lengths greater than 250 miles. Progressives are more common during the summer as they require more convection at the site of the frontal boundary.

Twelve parameters were collected in hourly intervals for all six stations beginning at 1200 UTC on June 29, 2012, until the time of the storm passage. The list of instability and wind shear parameters investigated are shown in **Table 2**. The instability parameters included the following: Convective Available Potential Energy (CAPE); the Most Unstable CAPE (MU-CAPE), the potential energy within the lowest 300 hPa; the Downdraft CAPE (DCAPE), measuring the strength of the rain-cooled downdraft; the 950–850 hPa lapse rate, describing lower-level instability; and convective temperature (CT), the surface temperature that must be attained for convection to occur. Precipitable water (PW) is the vertically integrated amount of water through the column of air. Wind shear parameters included the Bulk Richardson Number (BRN), which is a dimensionless ratio of turbulence versus wind shear. Helicity is a measure of the helical or "corkscrew" flow of the air. Additionally the 1 km storm relative inflow and three different layers of wind shear were included.

The model data were analyzed in BUFKIT [23], a visualization software designed for weather forecasting and analysis. The data were standardized by hour (...t-2, t-1, t) until the time of storm passage. Downshear Convective Available Potential Energy values were calculated by estimating the temperature of the downdraft parcel (T_{pd}), which is between the mid-level wet bulb potential temperature and the updraft wet bulb potential temperature [24] and is shown by the following equation:

$$\frac{1}{2} * g * \left(\frac{Te - Tpd}{Te} \right) * \Delta z \tag{1}$$

where T_e is the environmental surface temperature (K), T_{pd} is the expected parcel downdraft surface temperature (K), and Δz is the depth of the negatively buoyant air (m).

One-sample one-tailed *t*-tests were then conducted (for each station and each variable) at hourly intervals relative to the storm passage in order to determine if parameter values were significantly larger than the mean daily values and to potentially develop critical forecasting thresholds. The number of samples varied at each station since they were measured hourly starting at the beginning of the day until relative to the storm passage; therefore, more data were available progressing eastward. Level 2 Radar, which has higher resolution than conventional Level 3 Radar and offers dual polarmetric data [25] was also analyzed in addition to surface and upper air plots.

4. Results

A quasi-stationary (QS) boundary was oriented west-east from Iowa to the Mid-Atlantic region during the morning hours on June 29, 2012 (**Figure 6**). Close inspection of the soundings and surface plots revealed warm temperatures and high dew points in the region at 1500 UTC (**Figures 7** and **8**). On the northern side of the boundary, temperatures were in the mid-70s to the low-80s (Fahrenheit) (21–28°C), while dew points were around 60°F (16°C). However, on the southern side of the boundary, temperatures were in the mid-80s (30°C) to around 90°F (32°C), while the dew points were close to 70°F (21°C). The reason the air was unusually warm and humid in the morning hours was because it was the result of a decaying mesoscale convective system the previous day. A small region of convection then formed over northern Iowa as a strong southerly flow from the nocturnal low-level jet provided ample moisture to the system indicated by the dew point equaling the temperature around 900 hPa. The sounding at Davenport (**Figure 9**) at 1200 UTC exhibited a classic "inverted-v" shape confining the potential energy close to the surface and preventing thunderstorms from developing too quickly.

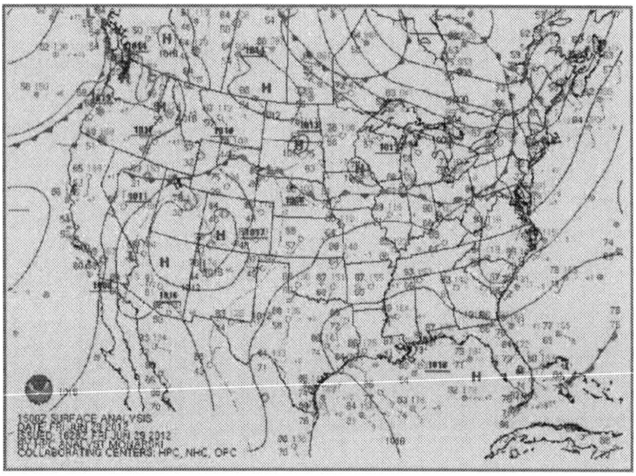

Figure 6. Surface map at 1500 UTC for June 29, 2012.

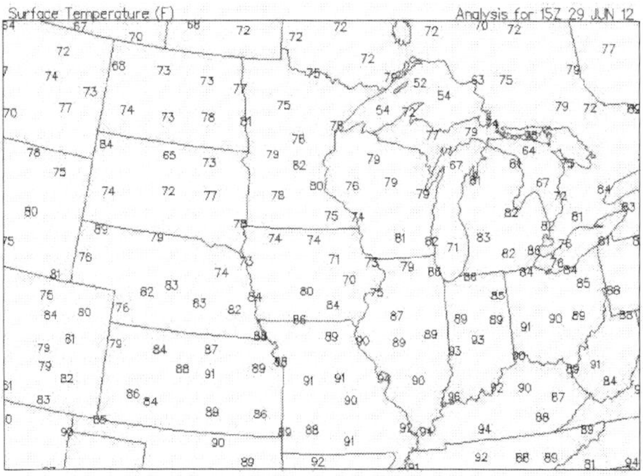

http://dx.doi.org/10.5772/63319

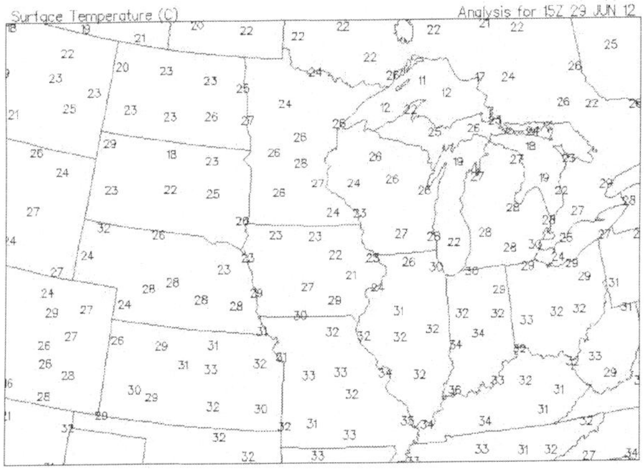

Figure 7. Surface plot of temperatures (a) (previous page °F); (b) (above °C) at 1500 UTC June 29, 2012.

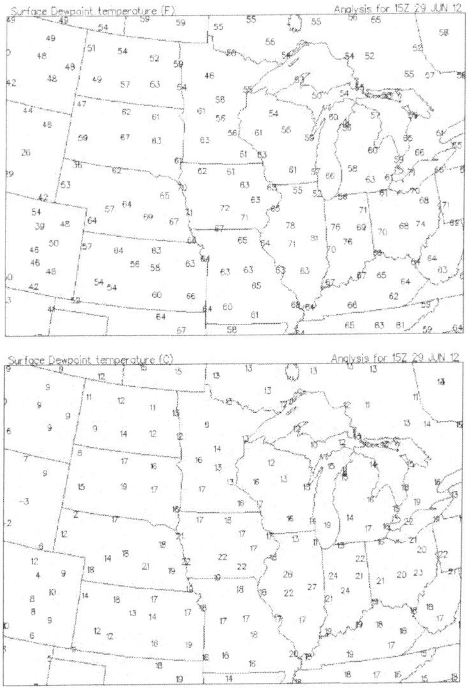

Figure 8. Surface plot of dew points (a) (top °F); (b) (bottom °C) at 1500 UTC for June 29, 2012.

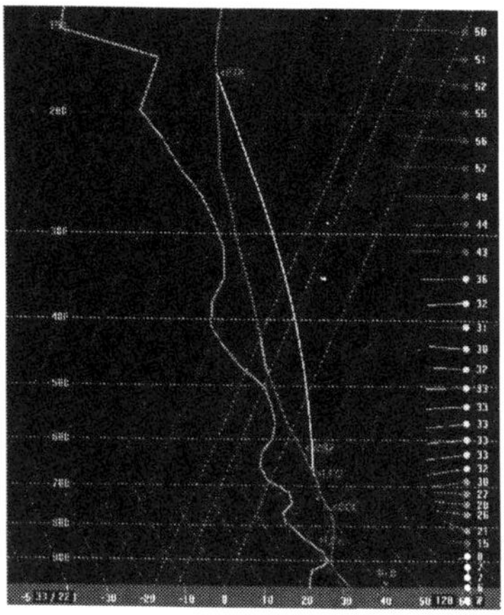

Figure 9. NAM 1200 UTC run for sounding at Davenport, IA for 1500 UTC, June 29, 2012.

A small of line of thunderstorms formed along the QS boundary and passed through Davenport, IA at approximately 1300 UTC. The 500 hPa chart did not show any short-wave troughs but exhibited strong westerly flow over northern Iowa (**Figure 10**). However, the 250 hPa chart showed a jet streak with the right entrance region over northern Iowa to provide synoptic scale forcing (**Figure 11**). Despite not having large CAPE values (less than 500 J kg^{-1}) in the early morning, the moderate amount of synoptic scale forcing was enough to develop the storms. By approximately 1500 UTC, the QS boundary was between Davenport and Chicago (**Figure 12**). The rear cold outflow combined with the southerly surface inflow appeared to "split" the system. In addition, the faster westerly mid–upper-level winds to the north of the boundary caused the eastern part of the QS boundary to turn clockwise into a north-south orientation (**Figure 13**).

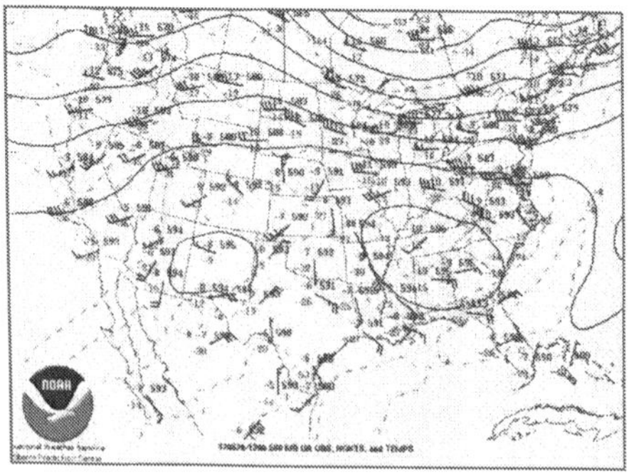

Figure 10. 500 hPa map at 1200 UTC, June 29, 2012.

the respective means of 2000 and 2200 J kg^{-1}. These significant results are largely due to the extremely high CAPE/MUCAPE model values on the order of 4500–5000 J kg^{-1} found in the region.

	t	Mean	SD	Test value
CT (°C)	*32.95	41.82	0.33	37
Lapse Rate (°C km^{-1})	-4.12	5.46	1.62	8.5
CAPE (J kg^{-1})	-32.74	65.6	132	2000
MUCAPE (J kg^{-1})	-2.75	1620	471	2200
DCAPE (J kg^{-1})	-6.9	618	122.8	1000
Shr (0–3) (m s^{-1})	*29.53	33.2	1.64	11.5
Shr (0–6) (m s^{-1})	*33.67	38.4	1.51	15.6
Shr (3–6) (m s^{-1})	-20.00	5.2	1.09	15
SR Inflow (m s^{-1})	*8.94	10.0	1.0	6
Helicity (m^2 s^2)	-0.10	148.4	36.08	150
PW (cm)	*83.98	3.71	0.03	2.54
BRN (unitless)	-103.16	0.52	1.07	50

*Significant parameter (CT, shr 0–3, shr 0–6, SR Inflow, PW).

Table 4. One-sample *t*-test for the mean daily values prior to derecho passage at Chicago, IL (df = 4, $p < .05$).

	t	Mean	SD	Test value
CT (°C)	0.13	37.13	3.14	37
Lapse Rate (°C km^{-1})	-0.82	8.06	1.69	8.5
CAPE (J kg^{-1})	*3.17	3288	1284.8	2000
MUCAPE (J kg^{-1})	*19.69	3988	272.5	2200
DCAPE (J kg^{-1})	-0.43	963	275	1000
Shr (0–3) (m s^{-1})	*5.00	21.9	6.59	11.5
Shr (0–6) (m s^{-1})	*9.05	34.3	6.53	15.6
Shr (3–6) (m s^{-1})	-0.87	10.7	15.6	15
SR Inflow (m s^{-1})	*9.00	8.7	0.95	6
Helicity (m^2 s^2)	-10.2	-19.7	52.38	150
PW (cm)	*20.51	4.34	0.28	2.54
BRN (unitless)	*2.45	90.14	51.74	50

*Significant parameter (CAPE, MUCAPE, shr 0–3, shr 0–6, SR Inflow, PW, BRN).

Table 5. One-sample *t*-test for the mean daily values prior to derecho passage at Wilmington, OH (df = 9, $p < .05$).

	t	Mean	SD	Test value
CT (°C)	*6.30	41.05	2.22	37
Lapse Rate (°C km^{-1})	0.66	8.85	1.89	8.5
CAPE (J kg^{-1})	*2.36	2852.9	1253.9	2000
MUCAPE (J kg^{-1})	*6.07	3383.3	674.6	2200
DCAPE (J kg^{-1})	-0.19	985.2	269.7	1000
Shr (0–3) (m s^{-1})	*4.57	23.3	8.97	11.5
Shr (0–6) (m s^{-1})	*8.77	38.1	8.87	15.6
Shr (3–6) (m s^{-1})	-0.86	14.6	1.67	15
SR Inflow (m s^{-1})	*3.39	8.0	2.04	6
Helicity (m^2 s^2)	-10.16	-3.41	52.29	150
PW (cm)	*12.20	4.06	0.43	2.54
BRN (unitless)	1.34	79.6	76.7	50

*Significant parameter (CT, CAPE, MUCAPE, shr 0–3, shr 0–6, SR Inflow, PW).

Table 6. One-sample t-test for the mean daily values prior to derecho passage at Charleston, WV (df = 11, $p < .05$).

In addition to the large amount of instability, the storm relative inflow was notably high once the derecho passed Fort Wayne, IN. In Wilmington, Charleston, and Washington DC, the mean storm relative inflow was significantly larger than the critical value of 6 m s^{-1} (**Tables 5** and **6**). The mean was at least 8 m s^{-1} in both Wilmington and Charleston and often approached 12 m s^{-1} in the late afternoon from the south. Overall, the mean Downdraft CAPE was not significantly larger than the critical value because the values were low in the morning which offset the higher values in the late afternoon. However, the DCAPE values were extremely high in the late afternoon (approaching 1300 J kg^{-1}) aiding the formation of the cold pool and strengthening the derecho.

5. Conclusions

Derechos have been known to cause similar types of damage to tornadoes in terms of monetary damage and fatalities. Even though they have straight-line winds rather than rotational, the environmental conditions prior and during derecho events are comparable as well. By investigating the intense 2012 Mid-Atlantic Derecho, the importance of the key thermodynamics and dynamics could be seen.

Anomalously high wind shear in the low- and mid-levels was shown to be vital in the propagation of the storms. Low-level mean wind shear was significantly larger at the 0–3 km layer (21.7 m s^{-1}) than the critical threshold of 11.5 m s^{-1} at the time of passage at all six stations in the path. Likewise the low-mid-level mean shear in the 0–6 km level was significantly larger (31.5 m s^{-1}) than the critical threshold of 15.6 m s^{-1} at the time of passage at all six stations.

Equally important, anomalously high heat values combined with great atmospheric instability (measured by CAPE/MUCAPE) were present at certain times during the outbreak. During the early development, CAPE/MUCAPE values were low suggesting they are not necessary as long as other synoptic forcing agents (i.e., jet streaks) are available. CAPE/MUCAPE values were not high in Illinois where the derecho formed; however, there was plenty of shear and synoptic forcing from the jet in northern Illinois to initialize the development.

Even though the synoptic uplift was absent from the lack of a jet streak in the Mid-Atlantic, surface temperatures reached the convective temperatures, triggering uplift into the unstable atmosphere. Afternoon temperatures in the area easily reached approximately 100°F (38°C). CAPE/MUCAPE model values were approximately 4500–5000 J kg^{-1} and model convective temperatures ranged between 38 and 40°C. On the other hand, the strength of the downdraft dynamics (DCAPE) was large but not found to be significantly larger.

In summary, CAPE/MUCAPE and the mean wind shear were significantly larger than their respective critical thresholds in the Mid-Atlantic. The 2012 Mid-Atlantic Derecho showed how the right combination of mean low-mid-level shear with jet stream dynamics caused it to initially develop while the intense heat led to the convective instability necessary to generate intensity. The event demonstrated that winds do not have to rotate to cause widespread damage and societal impacts and, as such, the term "derecho" was brought back to the forefront in the meteorological vocabulary.

Author details

Kevin Law

Address all correspondence to: law14@marshall.edu

Marshall University Huntington, West Virginia, USA

References

[1] Goliger, A. M., and R. V. Milford, 1998: A review of worldwide occurrence of tornadoes. *J. Wind Eng. Ind. Aerodyn.*, 74–76, 111–121.

[2] Hinrichs, G., 1888: Tornadoes and derechos. *Amer. Meteor. J.*, 5, 341–349.

[3] Johns, R. H., and W. D. Hirt, 1987: Derechos: Widespread convectively induced windstorms. *Wea. Forecasting*, 2, 32–49.

[4] Ashley, W. S., and T. L. Mote, 2005: Derecho hazards in the United States. *Bull. Amer. Soc.*, 86, 1577–1592.

[5] Gatzen, C., 2004: A derecho in Europe: Berlin, 10 July 2002. *Wea. Forecasting*, 19, 639–645.

[6] López, J. M., and J. Arús, 2004: A Mediterranean derecho: Catalonia (Spain), 17th August 2003. Preprints, *Third European Conf. on Severe Storms*, León, Spain, CD-ROM, P53.

[7] Fujita, T. T., and R. M. Wakimoto, 1981: Five scales of airflow associated with a series of downbursts on 16 July 1980. *Mon. Wea. Rev.*, 109, 1438–1456.

[8] Zipser, K. A., 1982: Use of a conceptual model of the life cycle of mesoscale convective systems to improve very short range forecasts. *Nowcasting*, K. Browning, Ed., Academic Press, London, 191–204.

[9] Bentley, M. L., and T. L. Mote, 1998: A climatology of derecho producing mesoscale convective systems in the central and eastern United States, 1986–1995. Part I: Temporal and spatial distribution. *Bull. Amer. Meteor. Soc.*, 79, 2527–2540.

[10] Fujita, T. T., 1978: Manual of downburst identification for project NIMROD, SMRP Res. Paper 156, University of Chicago, NTIS Accession No. N78-30771/7GI, 104 pp.

[11] Coniglio, M. C., and D. J. Stensrud, 2001: Simulation of a progressive derecho using composite initial conditions. *Mon. Wea. Rev.*, 129, 1593–1616.

[12] Bentley, M. L., and J. A. Sparks, 2003: A 15-yr climatology of derecho-producing mesoscale convective systems over the central and eastern United States. *Climate Res.*, 24, 129–139.

[13] Burke P. C., and D. M. Shultz, 2004: A 4-yr climatology of cold season bow echoes over the continental United States. *Wea. Forecasting*, 19, 1061–1074.

[14] Changnon S. A., 2001: Damaging thunderstorm activity in the United States. *Bull. Amer. Meteor. Soc.*, 82, 597–608.

[15] Brooks, H. E., and C. A. Doswell III, 2001: Normalized damage from major tornadoes in the United States: 1980–1999. *Wea. Forecasting*, 16, 168–176.

[16] National Weather Service, 2013: The historic derecho of June 29, 2012. Service Assessment, U.S. Department of Commerce, National Oceanic and Atmospheric Administration, Silver Spring, Maryland, 61 pp.

[17] The Weather Channel, Derecho: The science behind widespread damaging winds. Retrieved from https://weather.com/news/news/derecho-explainer-20120612.

[18] Accuweather, intense storms called derechos slam 700 miles of the US. Retrieved from http://www.accuweather.com/en/weather-news/deadly-super-derecho-strikes-m/67383.

[19] Coniglio, M. C., S. F. Corfidi, and J. S. Kain, 2011: Environment and early evolution of the 8 May 2009 derecho producing convective system. *Mon. Wea. Rev.*, 139, No. 4, 1083–1102.

[20] Duke, J. W., and J. A. Rogash, 1992: Multiscale review of the development and early evolution of the 9 April 1991 derecho. *Wea. Forecasting*, 7, 623–635.

[21] Evans, J. S., and C. A. Doswell, 2001: Examination of derecho environments using proximity soundings. *Wea. Forecasting*, 16, 329–342.

[22] Metz, N., and L. Bosart, 2010: Derecho and MCS development, evolution, and multi-scale interactions during 3–5 July 2003. *Mon. Wea. Rev.*, 138, 3048–3070.

[23] Warning Decision Training Division, National Weather Service, 2015. Retrieved from http://www. wdtb.noaa.gov/tools/BUFKIT/index.html.

[24] Gilmore M. S., and L. J. Wicker, 1998: The influence of midtropospheric dryness on supercell morphology and evolution. *Mon. Wea. Rev.*, 126, 943–958.

[25] Vasquez T, 2013: Weather radar handbook. WeatherGraphics, Garland, Texas, 152 pp.

Chapter 5

Extreme Precipitation Events over East Asia: Evaluating the CMIP5 Model

Nicolas Freychet, Huang-Hsiung Hsu and Chi-Hua Wu

Additional information is available at the end of the chapter

http://dx.doi.org/10.5772/62996

Abstract

Extreme hydrological events are a direct threat to society and the environment, and their study within the framework of global climate change remains crucial. However, forecasts present numerous uncertainties.

This study investigates the modification of precipitation characteristics over East Asia, a region densely populated and vulnerable to extreme rainfall. The performance of the models and the confidence in their projections are analyzed using data derived from an ensemble of models from Phase 5 of the Coupled Model Intercomparison Project (CMIP5). Different factors that can affect the confidence of the projection are considered, such as the resolution, the response to radiative forcing, and the modification of large-scale atmospheric circulation.

The resolution and response in radiative forcing do not exhibit a clear correlation with the change in precipitation. The moisture flux convergence (MFC), by contrast, has a clear impact on extreme events. Specifically, the change in the dynamical term of the vertical MFC exhibits a major disagreement among the models and could strongly affect the confidence of the ensemble projection. Extreme precipitation is likely to increase over East Asia and India.

Keywords: Extreme precipitation, CMIP5, East Asia monsoon, RCP8.5, variability

1. Introduction

The Asian subcontinent is one of the most densely populated areas on the Earth and is also one of the most vulnerable regions to environmental conditions. The region is subject to strong seasonal variations in precipitation, with wet summers and drier winters. The dynamics of the Asian summer monsoon have been reviewed in many papers [e.g., 1–5]. The monsoon provides necessary water for societal needs, but can also be associated with extreme precipitation. This type of rainfall represents a dual threat to society. First, the rain falls in excess amounts over a short period, resulting in a high runoff rather than recharging the groundwater supply. Second, extreme precipitation often leads to floods and, eventually, to landslides, building collapses, and causalities. Extreme precipitation in East Asia is often associated with typhoons, but it can also be triggered by other atmospheric conditions, such as a recent extreme rainfall event that occurred over China in the late spring and early summer of 2015. Heavy rainfall (above 60 mm day^{-1}) first affected the southern provinces before moving toward the central and northern parts of the country. Overall, several hundreds of thousands of people were affected, tens of thousands of houses were destroyed, several people died, and crops and roads were severely damaged.[1] These effects showcase the need for anticipating and planning hydrological extreme events, including answering the following question: How will global climate change affect the characteristics of extreme rainfall?

Various studies have investigated the projections of precipitation in a warming climate [6–19] and the modification of monsoon systems under such conditions [20–27]. The global and regional patterns of extreme rainfall changes have also been subject to research [28–36]. The Intergovernmental Panel on Climate Change Fourth Assessment Report (IPCCAR4) presents a summary of these studies; for example, Chapters 10.3.6 [37] and 11.4 [38] detail projections for Asia. However, one critical problem remains: To what extent can we trust the models? Phase 5 of the Coupled Model Intercomparison Project (CMIP5) provides a wide range of models, each with its own characteristics (e.g., parametrization and resolution). The study of an ensemble extracted from the CMIP5 is a good approach to investigating the uncertainties associated with the projection and to quantifying its confidence. In this study, 23 models are used, all in accordance with the representative concentration pathway 8.5 (RCP8.5), to answer the following questions: How will precipitation characteristics change over East Asia by the end of the twenty-first century? How reliable are these projections? What can explain the scattering of the ensemble? Moreover, what will be the social and economic impacts?

Section 2 details the data and methodology used. Section 3 introduces a review of the performance of the CMIP5 models as well as a comparison against observations and then an analysis on the estimated ensemble projection and its reliability. Section 4 presents a more detailed discussion on the change in atmospheric characteristics associated with the change in extreme rainfall. Section 5 provides a brief discussion on possible societal impacts, and Section 6 offers a conclusion.

[1] Statistics from: floodlist.com/tag/china

http://dx.doi.org/10.5772/62996

2. Data and methodology

2.1. Data

To examine the projection and reliability of the CMIP5 ensemble, 23 models (**Table 1**) are used (with a single member for each). The models were first forced by historical conditions (the observation of aerosols, greenhouse gases, and solar irradiance) until 2005 and then followed the RCP8.5 pathway [39], leading to an increase in radiative forcing of 8.5 W m^{-2} by the end of 2100. The projection of the ensemble is analyzed by conducting a comparison between the last 30 years of the century (i.e., from 2071 to 2100, called "RCP" hereafter) and 30 years of the historical period (i.e., from 1976 to 2005, called "HIST" hereafter). Extreme events are computed from the daily outputs, and then the results are averaged for the 30 years of each period.

Model Name	Institute and Country	Resolution
ACCESS1.0	Commonwealth Scientific and Industrial Research Organization (CSIRO), and Bureau of Meteorology (BoM), Australia	144 × 192
ACCESS1.3	CSIRO and BoM, Australia	144 × 192
BCC-CSM1.1	Beijing Climate Center (BCC) and China Meteorological Administration (CMA), China	64 × 128
BCC-CSM1.1(m)	BCC and CMA, China	160 × 320
BNU-ESM	Beijing Normal University (BNU)–Earth System Model, China	64 × 128
CanESM2	Canadian Centre for Climate Modeling and Analysis (CCCma), Canada	64 × 128
CMCC-CESM	Centro Euro-Mediterraneo per i Cambiamenti Climatici (CMCC), Italy	48 × 96
CMCC-CM	CMCC-CM, Italy	240 × 480
CMCC-CMS	CMCC-CM, Italy	96 × 192
CNRM-CM5	Centre National de Recherches Météorologiques (CNRM), and Centre Européen de Recherches et de Formation Avancée en Calcul Scientifique, France	128 × 256
CSIRO-Mk3–6-0	CSIRO Marine and Atmospheric Research (Melbourne) in collaboration with the Queensland Climate Change Centre of Excellence (QCCCE) (Brisbane)	96 × 192
FGOALS-G2	Institute of Atmospheric Physics, Chinese Academy of Sciences (IAP), and Tsinghua University (THU), China	60 × 128
GFDL-CM3	Geophysical Fluid Dynamics Laboratory (GFDL), USA	90 × 144
GFDL-ESM2G	GFDL, USA	90 × 144
GFDL-ESM2M	GFDL, USA	90 × 144
IPSL-CM5A-LR	Institut Pierre-Simon Laplace (IPSL), France	96 × 96
IPSL-CM5A-MR	IPSL, France	143 × 144

Model Name	Institute and Country	Resolution
IPSL-CM5B-LR	IPSL, France	96 × 96
MIROC5	Atmosphere and Ocean Research Institute (AORI; The University of Tokyo), National Institute for Environmental Studies (NIES), and Japan Agency for Marine-Earth Science and Technology (JAMSTEC), Japan	128 × 256
MIROC5-ESM-CHEM	JAMSTEC, AORI, and NIES, Japan	64 × 128
MPI-ESM-LR	Max Planck Institute for Meteorology (MPI-M), Germany	96 × 192
MRI-CGCM3	Meteorological Research Institute, Japan	160 × 320
NorESM1-M	Norwegian Climate Centre, Norway	96 × 144

Table 1. CMIP5 models used for this study. The resolution is indicated in grid points (latitude × longitude).

Asia covers a large area with different regional climatologies (**Figure 1**). The mean precipitation during the extended summer (May–August) season is computed from the daily (1997–2007) Global Precipitation Climatology Project (GPCP) observations [40]. The total domain is subdivided into three regions based on the different phases of the Asian summer monsoon: East Asian region (EAR: 22°–45°N, 105°–145°E), India region (IR: 5°–28°N, 70°–105°E), and the western North-Pacific region (WNPR: 5°–22°N, 105°–160°E). The results are averaged over each of the three subregions to compare the regional characteristics of the rainfall.

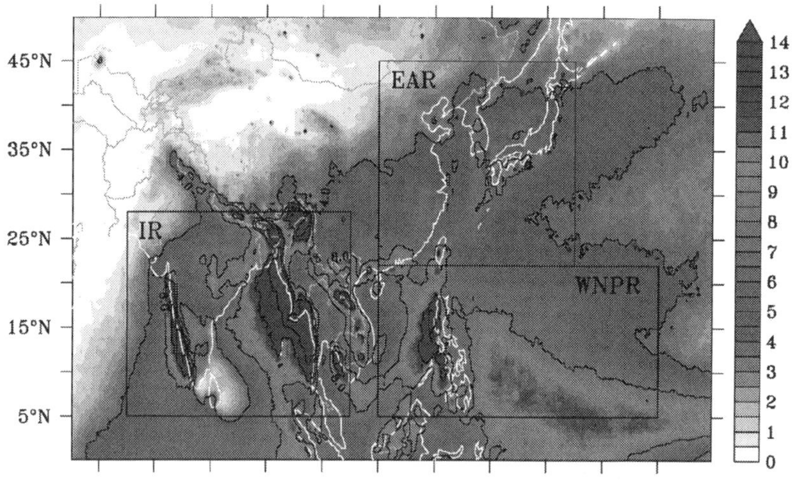

Figure 1. Mean May–August precipitation (mm day^{-1}) from GPCP [40] daily data (1997–2007), and an illustration of the three main domains used for this study: EAR(22°–45°N, 105°–145°E), IR(5°–28°N, 70°–105°E), and WNPR (5°–22°N, 105°–160°E).

2.2. Methodology

The overall methodology of this study follows that used in [41]. To study the characteristics of precipitation, the probability density function (pdf) is computed to separate and highlight the types of rainfall. Both intensity and frequency are considered and are defined as follows.

2.2.1. *Intensity*

Intensity is computed by separating the precipitation into percentiles at each grid point. Values are computed by steps of 10 between 1 and 90 and by steps of 1 between 90 and 99. Percentiles are computed either for the entire year (using all 30 years) or for each month (using only the precipitation from identical months over 30 years). The intensity of a percentile X is denoted as *pctX* and is expressed in mm day^{-1}.

2.2.2. *Frequency*

Frequency indicates the occurrence of a certain range of precipitation. It is computed for each grid point by using a threshold based on the HIST intensity value of a yearly percentile (this threshold is therefore independent from the month) and averaged over a 10° × 10° box around this point. At each grid point, each day with rainfall higher than this threshold is counted as an event (for the given percentile). The HIST and RCP use the same threshold value for a percentile, and thus the same definition for light and heavy rainfall is used for both periods, which facilitates comparing the change in each percentile. The frequency is determined by averaging the count of events from the same months over 30 years. The frequency of a percentile X is denoted as *fqpctX* and is expressed as a number of days.

The projection is expressed either as a difference between the two periods (RCP – HIST) or as a relative difference ([(RCP – HIST)/HIST] × 100). The former has a unit depending on the variable, whereas the relative difference is always expressed as a percentage. Each value was computed independently for every model on the respective grids. To plot the spatial distributions, the results were interpolated on a common grid (1.5° × 1.5°) to compute the ensemble mean and the associated ensemble standard deviation.

A change in intensity translates how each type of rainfall (light, medium, and heavy) could become more or less intense, while the change in frequency indicates how each type of rainfall could become more or less frequent.

3. Confidence in the distribution of precipitation and its projection

3.1. A brief review of the historical CMIP5 ensemble performances

First, the capability of the ensemble to simulate the precipitation and monsoon circulation over Asia is reviewed briefly. **Figure 2** shows the annual signal of precipitation in Asia for the 1976–2005 period (HIST) for the ensemble mean and the GPCP daily data (averaged over 1997–2007). The 99th percentile is also computed for each grid point over the entire period and then

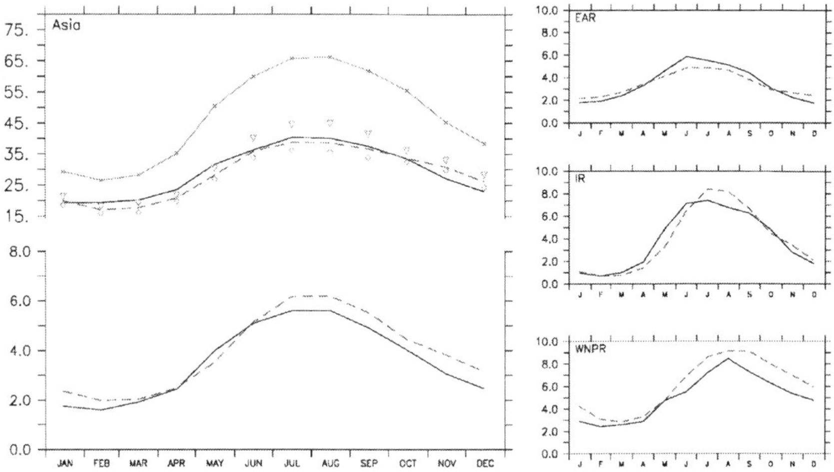

Figure 2. Annual signal of precipitation (mm day^{-1}) in the Asia region (5°–45°N, 65°–160°E), EAR (22°–45°N, 105°–145°E), IR (5°–28°N, 70°–105°E), and WNPR (5°–22°N, 105°–160°E). The solid black line represents the GPCP observations, and the dashed red line indicates the CMIP5 ensemble mean. For the Asia region (left), on the top panel, the solid black and dashed red lines indicate the 99th percentile for the GPCP and CMIP5, respectively, and the orange triangles (circles) correspond to the mean value of the 10 models with the highest (lowest) resolution. The PERSIANN observational dataset is also plotted for the 99th percentile (solid green line). The signal is averaged over 1976–2005 for the CMIP5 ensemble and 1997–2007 for the observations.

averaged for the Asia domain (5°–45°N, 65°–160°E). Another observational dataset is used for the 99th percentile (PERSIANN, [42]), which has a 0.25° resolution.

The mean signal of the ensemble is relatively consistent with observations, and it can detect the seasonal signal for each region, although it tends to overestimate (underestimate) the precipitation over WNPR (EAR) during the summer. A significant delay in the onset of the monsoon over IR is also visible. The signal of extreme precipitation (99th percentile) is also adequately represented in a comparison between the CMIP5 mean and the GPCP. However, the signal intensity is stronger with a higher-resolution dataset (PERSIANN). This impact of the resolution could also affect the individual model results. Thus, for the 99th percentile, we separated the 10 models with the highest resolution (triangles) and the 10 models with the lowest resolution (circles). The effect of the resolution is clearly apparent, establishing an increased estimation of the 99th intensity according to the highest-resolution models during the Asian summer monsoon (the triangles are closer to the observation line.).

Regarding the spatial distribution (**Figure 3**) for the mean precipitation and wind circulation, biases are more readily apparent. In this figure, two periods are separated to highlight the monsoon evolution: May–June (MJ) and July–August (JA). Overall, the mean circulation is adequately represented by the CMIP5 ensemble, as are the rainfall patterns, although the intensity of precipitation exhibits a low bias more clearly over the Mei-Yu front in the EAR (during MJ) and a negative (positive) bias over the northern (southern) part of the IR, partic-

The correlations are also more apparent in **Figure 12** and **Table 3**, particularly for the total change in the VMFC. The changes in extreme precipitation characteristics are strongly correlated to the change in the VMFC (0.50–0.89). When considered separately, the correlation between the individual terms (VMFCt and VMFCd) with the changes in extreme precipitation become less clear. This is particularly applicable to the EAR, where significant correlations are found only for the total VMFC. The VMFCt significantly affects the WNPR in both frequency (0.75) and intensity (0.61), and the VMFCd is correlated with the frequency of extreme precipitation over the IR (0.63). These results clearly revealed that the change in extreme precipitation, both in frequency and intensity, is strongly affected by the change in the VMFC, and mostly under the simultaneous consideration of thermodynamic and dynamic terms.

	Intensity			Frequency		
	VMFC	VMFCt	VMFCd	VMFC	VMFCt	VMFCd
IR	**0.50**	0.12	0.33	**0.89**	0.41	**0.63**
WNPR	**0.67**	**0.61**	0.46	**0.72**	**0.75**	0.46
EAR	**0.67**	0.3	0.02	**0.51**	0.26	−0.14

Bold text indicates values exceeding the 99% confidence level.

Table 3. Identical to **Table 2**, except for vertical moisture flux convergence (VMFC) and its thermodynamical (VMFCt) and dynamical (VMFCd) components.

As mentioned in [41], these results may at first be contrary to expectations, which might anticipate an increase in atmospheric humidity content to be the dominant factor for explaining the change in extreme precipitation. The present research shows that the dynamical part of the VMFC can also significantly influence the intensity and frequency of these events. Furthermore, because its change is less apparent (**Figure 11**), it negatively influences confidence levels in projecting extremes (by increasing the scattering of the ensemble). By contrast, the change in the VMFCt is clearer and tends to increase confidence levels in projecting extremes.

4.4. Summary

In a warming climate, the specific humidity of the atmosphere is typically expected to increase. The variation in monsoon areas and precipitation has been linked to a rise in moisture convergence, offset by the change in circulation [15]. The earlier sections showed that the VMFC exerts a potentially high level of influence on the trends of extreme events, whereas the correlation between the change in the mean circulation and the HMFC is less clear. However, the influence of the VMFC varies significantly between regions, which is consistent with findings in [46], showing that the dynamic contribution induces spatial variations in the changes. The confidence of ensemble-projected change in extreme precipitation may clearly be influenced by the low level of agreement in the model dynamics, particularly in the IR and WNPR. A direct correlation was not found between the resolution of the models or their response to radiative forcing and the change in extreme events, at least with the ensembles used for this study.

5. Extreme precipitation projections and societal impacts

These findings highlight the importance of considering regional studies when examining long-term projections of precipitation characteristics. Because of the different responses of the models to radiative forcing (in atmospheric circulation), the change in precipitation can be uncertain. However, the clearest and strongest tendencies are observed for heavy and extreme precipitation, which is also the most critical in terms of constituting a threat to society (but not for water resources, which is another subject of concern). Moreover, the regions for which the changes are most significant for these events are also those that are the most densely populated: East Asia and South Asia. Over these two regions, heavy rainfall is expected to increase in frequency and intensity with good confidence.

Considering past events (such as the floods of 2011, 2013, or 2015 that affected South Asia and China) and their cost, extreme hydrological events represent a severe threat to the populations living in these regions [47, 48]. It is most likely that the loss associated with these events (in terms of economy, infrastructure, land, and casualties) will increase by the end of the century. However, even if an increase in heavy rainfall cannot be avoided, its impact can be limited [49]. For instance, because deforestation is a major cause of landslides under heavy rainfall [50], improved management over forests could locally limit the impact of heavy rainfall [51]. In urban areas, drainage systems and infiltration surfaces can be improved to convey more surface water, as indicated in [52], which compared measurements conducted in various cities. Thus, if measures are considered and implemented sufficiently early, it is possible to limit the impact of heavy rainfall and its associated costs.

6. Discussion and concluding remarks

The change in extreme precipitation under global warming conditions will remain a major subject of research for the next few decades. Extreme precipitation events can drastically affect society, entailing severe economic, infrastructure, and human loss. The use of a model ensemble is a beneficial approach to highlighting regions where a change in these events is clear or, alternatively, regions where the level of agreement is lower. The frequency of extreme precipitation and the confidence level of the projected changes are critical considerations for decision makers in developing the prevention and risk management measures.

This study used 23 models from the CMIP5. Certain projections of precipitation characteristics can be clearly identified as confident, whereas other results remain uncertain (**Figure 5**). The selected models revealed a high confidence level in indicating that the East Asian region will be highly vulnerable to global warming, exhibiting more intense and more frequent extreme precipitation. The India and South Asian regions exhibit the same tendency but with less confidence. The results in the western North Pacific region are less certain, particularly regarding the intensity of extremes. These findings are in agreement with those presented in [41], even if the ensemble used was different. Changes in the precipitation characteristics are likely to affect all Asian regions, but not with the same magnitude.

Different factors that can affect the scattering of the ensemble projection were studied through correlation analysis. The resolution of the models does not display a clear impact, although we found that it can significantly affect the capacity of the models to produce heavy rainfall (**Figure 4**). The response of the models to radiative forcing did not exhibit a clear relationship with the change of extremes (**Figure 6**). The most significant findings were obtained by considering daily moisture convergence over the entire atmosphere (Figures 9–12 and Tables 2 and 3). The vertically integrated moisture gradient can strongly influence the projection of extreme events. Specifically, the model results were not in agreement with the change of its dynamical term (VMFCd), and thus this may explain the uncertainties associated with the projection. Thus, even if the increase in atmospheric moisture content is expected to strengthen the precipitation, local variations in the circulation will strongly affect changes in the characteristics of extreme events.

The systematic bias of the CMIP5 models mentioned in Section 3, their low-level ability to simulate tropical cyclones correctly (with a resolution coarser than 1°), and the definition of the MF adopted herein could also lead to a biased estimation of the correlations computed for this study. Thus, the results should be considered indicators of what can affect the projection of extreme precipitation and ensemble uncertainties. However, other factors could also influence the response of the models and should be investigated in future studies. Nevertheless, most of the results indicate a confident increase in extreme precipitation over East Asia. The societal and economic impacts are expected to be massive, and anticipation will be critical to limiting the damage of such changes in hydrological events.

Acknowledgements

This work was supported by the Consortium for Climate Change Study (CCliCS) - Ministry of Science and Technology (MOST), Taiwan, under Grant MOST 100-2119-M-001-029-MY5. We thank the reviewers and editor for comments that greatly improved the manuscript.

Author details

Nicolas Freychet*, Huang-Hsiung Hsu and Chi-Hua Wu

*Address all correspondence to: nfreychet@gate.sinica.edu.tw

Research Center for Environmental Changes, Taipei, Taiwan

References

[1] Ramage CS. Monsoon Meteorology, International Geophysical Series, Vol. 15. Academic Press; 1971, San Diego. 296 p.

[2] Ding Y. The variability of the Asian summer monsoon. In: Kluwer Academic, editor. Monsoon over China. J. Meteor. Soc. Japan; 1994, pp. 85B, 21-54. DOI: 10.2151/jmsj.85B.21

[3] Chang C-P, Wang B, Lau N-C G. The global monsoon system: research and forecast. WMO/TD 1266, TMPRRep. 70 ed. 2005. 542 p.

[4] Wang B. The Asian Monsoon. Praxis Publishing; 2006, Chichester, UK. 787 p.

[5] Ding Y. The variability of the Asian summer monsoon. J. Meteor. Soc. Japan. 2007;85B: 21-54. DOI: 10.2151/jmsj.85B.21

[6] Chou C, Neelin J D. Mechanism of global warming impacts on regional tropical precipitation. J. Climate. 2004;17:2688–2701. DOI: 10.1175/1520-0442(2004)017,2688:MOGWIO.2.0.CO;2

[7] Min S-K, Legutke S, Hense A, Cubasch U, Kwon W-T, Oh J-H, Schlese U.. East Asian climate change in the 21st century as simulated by the coupled climate model ECHO-G under IPCC SRES scenarios. J. Meteor. Soc. Japan. 2006;84:1-26. DOI: 10.2151/jmsj.84.1

[8] Kripalani RH, Oh J-H, Chaudhari HS. Response of the East Asian summer monsoon to doubled atmospheric CO2: coupled climate model simulations and projections under IPCC AR4. Theor. Appl. Climatol. 2007;87:1-28. DOI: 10.1007/s00704-006-0238-4

[9] Stephens GL, Ellis TD. Controls of global-mean precipitation increases in global warming GCM experiments. J. Climate. 2008;21:6141-6155. DOI: 10.1175/2008JCLI2144.1

[10] Chou C, Neelin J D, Chen C-A, Tu J-Y. Evaluating the "rich-get-richer" mechanism in tropical precipitation change under global warming. J. Climate. 2009;22:1982–2005. DOI: 10.1175/2008JCLI2471.1

[11] Wu C-H, Kau W-S, Chou M-D. Summer monsoon onset in the subtropical western North Pacific. Geophys. Res. Lett. 2009;36:L18810. DOI: 10.1029/2009GL040168

[12] Seager R, Naik N, Vecchi GA. Thermodynamic and dynamic mechanisms for large-scale changes in the hydrological cycle in response to global warming. J. Climate. 2010;23:4651-4668. DOI: 10.1175/2010JCLI3655.1

[13] Chou C, Chen C-A, Tan P-H, Chen K-T. Mechanisms for global warming impacts on precipitation frequency and intensity. J. Climate. 2012;25:3291-3306. DOI: 10.1175/JCLI-D-11-00239.1

[14] Kusunoki S, Arakawa O. Change in the precipitation intensity of the East Asian summer monsoon projected by CMIP3 models. Climate Dyn. 2012;38:2055-2072. DOI: 10.1007/s00382-011-1234-7

[15] Hsu P-C, Li T, Luo J-J, Murakami H, Kitoh A, Zhao M. Increase of global monsoon area and precipitation under warming: a robust signal? Geophys Res Lett. 2012;39:L06701. DOI: 10.1029/2012GL051037

[16] Ma J, Xie S-P, Kosaka Y. Mechanisms for tropical tropospheric circulation change in response to global warming. J Climate. 2012;25:2979-2994. DOI: 10.1175/JCLI-D-11-00048.1

[17] Wu C-H, Chou M-D. Upper-tropospheric forcing on late-July monsoon transition in East Asia and western North Pacific. J Climate. 2012;25:3929-3941. DOI: 10.1175/JCLI-D-11-00343.1

[18] Seo, K-H, Ok J, Son J-H, Cha D-H. Assessing future change in the East Asian summer monsoon using CMIP5 coupled models. J Climate. 2013;26:7662-7675. DOI: 10.1175/JCLI-D-12-00694.1

[19] Lee J-Y, Wang B. Future change of global monsoon in the CMIP5. Climate Dyn. 2014;42:101-119. DOI: 10.1007/s00382-012-1564-0

[20] Wang B, Ding Q. Changes in global monsoon precipitation over the past 56 years. Geophys Res Lett. 2006;33:L06711. DOI: 10.1029/2005GL025347

[21] Inoue T, Ueda H. Delay of the first transition of Asian summer monsoon under global warming condition. SOLA. 2011;7:081-084. DOI: 10.2151/sola.2011-021

[22] Min S-K, Legutke S, Hense A, Cubasch U, Kwon W-T, Oh J-H, Schlese U. Projected changes in Asian summer monsoon in RCP scenarios of CMIP5. Atmos. Oceanic Sci Lett. 2012;5:43-48.

[23] Turner AG, Annamalai H. Climate change and the South Asian summer monsoon. Nat Climate Change. 2012;2:587-595. DOI: 10.1038/nclimate1495

[24] Duan A, Hu J, Xiao Z.. The Tibetan Plateau summer monsoon in the CMIP5 simulations. J Climate. 2013;26:7747-7766. DOI: 10.1175/JCLI-D-12-00685.1

[25] Hsu P-C, Li T, Luo J-J, Murakami H, Kitoh A, Zhao M. Increase of global monsoon area and precipitation under global warming: A robust signal? Geophys Res Lett. 2012;39:L06701. DOI: 10.1029/2012GL051037

[26] Jones C, Carvalho LMV. Climate change in the South American monsoon system: Present climate and CMIP5 projections. J Climate. 2013;26:6660-6678. DOI: 10.1175/JCLI-D-12-00412.1

[27] Seth A, Rauscher SA, Biasutti M, Giannini A, Camargo SJ, Rojas M. CMIP5 projected changes in the annual cycle of precipitation in monsoon regions. J Climate. 2013;26:7328-7351. DOI: 10.1175/JCLI-D-12-00726.1

[28] Trenberth KE, Dai A, Rasmussen R, Parsons D. The changing character of precipitation. Bull Am Meteor Soc. 2003;84:1205-1217. DOI: 10.1175/BAMS-84-9-1205

[29] Kharin VV, Zwiers FW. Estimating extremes in transient climate change simulations. J Climate. 2005;18:1156-1173. DOI: 10.1175/JCLI3320.1

[30] Meehl GA, Arblaster JM, Tebaldi C. Understanding future patterns of increased precipitation intensity in climate model simulations. Geophys Res Lett. 2005;32:L18719. DOI: 10.1029/2005GL023680

[31] Räisänen J. Impact of increasing CO2 on monthly-to-annual precipitation extremes: analysis of the CMIP2 experiments. Climate Dyn. 2005;24:309-323. DOI: 10.1007/s00382-004-0510-1

[32] Barnett DN, Brown SJ, Murphy JM, Sexton DMH, Webb MJ. Quantifying uncertainty in changes in extreme event frequency in response to doubled CO2 using a large ensemble of GCM simulations. Climate Dyn. 2006;26:489-511. DOI: 10.1007/s00382-005-0097-1

[33] Tebaldi C, Hayhoe K, Arblaster M, Meehl GA. Going to the extremes. Climatic Change. 2006;79:185-211. DOI: 10.1007/s10584-006-9051-4

[34] Giorgi F, Im E-S, Coppola E, Diffenbaugh NS, Gao XJ, Mariotti L, Shi Y. Higher hydroclimatic intensity with global warming. J Climate. 2011;24:5309-5324. DOI: 10.1175/2011JCLI3979.1

[35] Shiu C-J, Liu S C, Fu C, Dai A, Sun Y. How much do precipitation extremes change in a warming climate?. Geophys Res Lett. 2012;39:L17707. DOI: 10.1029/2012GL052762

[36] Scoccimarro E, Gualdi S, Bellucci A, Zampiery M, A. Navara A. Heavy precipitation events in a warmer climate: Results from CMIP5 models. J Climate. 2013;26:7902-7911. DOI: 10.1175/JCLI-D-12-00850.1

[37] Meehl GA, et al. Global climate projections. In: Climate Change 2007, The Physical Science Basis. S. Solomon, et al. Eds., Cambridge, UK, Cambridge University Press; 2007. pp. 747–845.

[38] Christensen JH, et al. Regional climate projections. In: Climate Change 2007, The Physical Science Basis. S. Solomon, et al. Eds., Cambridge, UK, Cambridge University Press; 2007. pp. 847–940.

[39] Riahi K, Gruebler A, Nakicenovic N. Scenarios of long-term socio-economic and environmental development under climate stabilization. Technol. Forecast Soc Change. 2007;74:887-935. DOI: 10.1016/j.techfore.2006.05.026

[40] Huffman G, Adler RF, Morrissey MM, Bolvin DT, Curtis S, Joyce R, McGavock B, Susskind J. Global precipitation at one-degree daily resolution from multisatellite observations. J Hydrometeor. 2001;2:36-50. DOI: 10.1175/1525-7541(2001)002,0036:GPAODD.2.0.CO;2

[41] Freychet N, Hsu H-H, Chou C, Wu C-H. Asian summer monsoon in the CMIP5 projections: A link between the change in extreme precipitation and monsoon dynamics. J Climate. 2015;28:1477-1493.

[42] Sorooshian S, Hsu KL, Gao X, Gupta HV, Imam B, Braithwaite D. Evaluation of PERSIANN system satellite based estimates of tropical rainfall. Bull Am Meteor Soc. 2000;81:2035-2046.

[43] Wu C-H, Hsu H-H, Chou M-D. Effect of the Arakan Mountains in the northwestern Indochina Peninsula on the late May Asian monsoon transition. J Geophys Res Atmos. 2014;119:10 769-10 779. DOI: 10.1002/2014JD022024

[44] Lee J-Y, Wang B. Future change of global monsoon in the CMIP5. Climate Dyn. 2014;42:101-119. DOI: 10.1007/s00382-012-1564-0

[45] Turner AG, Slingo JM. Uncertainties in future projections of extreme precipitation in the Indian monsoon region. Atmos Sci Lett. 2009;10:152-158. DOI: 10.1002/asl.223

[46] Chen C-A, Chou C, Chen C-T. Regional perspective on mechanisms for tropical precipitation frequency and intensity under global warming. J Climate. 2012;25:8487-8501. DOI: 10.1175/JCLI-D-12-00096.1

[47] Lin W, Gu W. The Eastern China flood of June 2015 and its causes. Sci Bull. 2016;61(2): 178-184.

[48] Promchote P, Wang S-YS, Johnson PG. The 2011 Great Flood in Thailand: climate diagnostics and implications from climate change. J Climate. 2016;29(1):367-379. DOI: http://dx.doi.org/10.1175/JCLI-D-15-0310.1

[49] Patankar AM, Patwardhan A, Marome WA, Porio E. Enhancing adaptation to climate change by integrating climate risk into long-term development plans and disaster management. APN E-Lib [Internet]. 2012. Available from: http://www.apn-gcr.org/resources/items/show/1572

[50] Chintraruck A, Walsh J. Bangkok and the Floods of 2011, Editor: M. Ann Miller and M. Douglass, Asia Research Institute, National University of Singapore: urban governance and the struggle for democratisation. In: Disaster Governance in Urbanising Asia. Springer, Singapore; 2016. pp. 195-209. DOI: 10.1007/978-981-287-649-2_10

[51] Lachica A P. Degraded forest rehabilitation and sustainable, forest management in East Asia region. In: APFNet Thematic Training Workshop on Degraded Forest, Rehabilitation and Sustainable Forest Management; 1–12 July 2014; 2014. pp. 79-87.

[52] Patwardhan A, Marome W. Characterizing public and private adaptation to climate change and implications for long-term adaptive capacity in Asian Megacities, APN E-Lib. 2016. Available from: http://www.apn-gcr.org/resources/items/show/1897

2.2. The dynamic Smagorinsky model

The unsolved scales can be viewed as the sub-grid scale viscosity due to the effect of small vortices whose size is less than Δ. The effect of small scales is to transfer the kinetic energy from large scales for dissipation following the theatrical energy cascade of *Kolmogorov*; then this behaviour can be modelled as a sub-grid stress (SGS) tensor $\tilde{\tau}_{ij}$, $\tilde{\tau}_{i\psi}$. Furthermore, while large-scale motions are strongly dependent on the external flow conditions, small-scale motions are expected to behave more universally. Hence, the intention is that numerical modelling can be feasible and/or require few adjustments when applied to various flows [19].

Eddy viscosity models parameterized the SGS stress tensor in a way to relate it with the resolved scales through the deformation tensor $S_{ij}^{\oplus}$:

$$\tau_{ij}^{a} = -2\nu_T S_{ij}^{\oplus} \tag{8}$$

where $\tau_{ij}^{a} = \tau_{ij} - {}^{1}\!/_{3}\delta_{ij}\tau_{kk}$ is the anisotropic part of SGS stress tensor, and ν_T denotes the SGS viscosity due to small scales.

The *Smagorinsky model* [16] proposes the equilibrium state where the small scales dissipate the kinetic energy entirely and instantaneously they receive from the large scales [20]. Following this assumption

$$\nu_T = C\Delta^2 \left|S^{\oplus}\right| \tag{9}$$

with $|S^{\oplus}| = \left(2S_{ij}^{\oplus}S_{ij}^{\oplus}\right)^{1/2}$ and $C_s = \sqrt{C}$ known as the *Smagorinsky coefficient*. The C_s coefficient takes values between 0.18 and 0.23 depending on the kinetic energy, Reynolds number, solid boundary proximity and time-space decay energy.

More sophisticated models allow to progressively estimate the *Smagorinsky coefficient* in time space. The *dynamic Smagorinsky model* introduces a test filter operation where the filter size is taken as $\Delta^T = \alpha\Delta$. Typically $\alpha = 2$ [17, 18]. The model coefficient is calculated to apply the *Germano identity*:

$$L_{ij}^{T} = T_{ij} - \tau_{ij}^{T} = \left(u_i^{\oplus}u_j^{\oplus}\right)^{T} - u_i^{\oplus T}u_j^{\oplus T} \tag{10}$$

where L_{ij}^{T} denotes the Leonard test filter stress, and T_{ij} denotes the SGS stress to the test filter scale. The last term can be modelled following the eddy viscosity model (Eq. (8)):

$$T_{ij} - \frac{1}{3}\delta_{ij}T_{kk} = -2C_T\left(\Delta^T\right)^2\left|S^{\oplus T}\right|S_{ij}^{\oplus T} \tag{11}$$

Based on Eqs. (8) and (11), the relation in Eq. (10) can be estimated as

$$\begin{aligned} L_{ij}^T &= -2C_T\Delta^2\underbrace{\left[\alpha^2\left|S^{\oplus}\right|^T S_{ij}^{\oplus T} - \left(\left|S^{\oplus}\right|S_{ij}^{\oplus}\right)^T\right]}_{-M_{ij}} \\ &= 2C_T\Delta^2 M_{ij} \end{aligned} \tag{12}$$

where the scale invariance of the coefficients ($C_T = C$) has been assumed, and C_T is found by optimizing through least squares minimization of error function $\varepsilon rr = (L_{ij}^T - 2C_T\Delta^2 M_{ij})^2$ when $\frac{\partial(\varepsilon rr)}{\partial C_T} = 0$ [17]:

$$C_T = \frac{1}{2\Delta^2}\frac{\left\langle L_{ij}^T M_{ij}\right\rangle_H}{\left\langle M_{ij}M_{ij}\right\rangle_H} \tag{13}$$

The brackets $\langle\, . \,\rangle_H$ in Eq. (13) denote the average over homogeneous directions and need to be introduced in order to guarantee numerical stability of the procedure because when $C_T < 0$, the numerical stability is lost. This operation creates an important limitation, restricting the simulations to flat terrains.

In order to avoid this constraint, a *modified dynamic Smagorinsky model* is proposed using another test filter size $\Delta^T = 3\Delta$ ($\alpha = 3$) and computing the test filter coefficient [8] as

$$\begin{cases} \dfrac{1}{2\Delta^2}\dfrac{L_{ij}^T M_{ij}}{M_{ij}M_{ij}} > 0 \quad \text{then} \quad C_T = \dfrac{1}{2\Delta^2}\dfrac{L_{ij}^T M_{ij}}{\left(M_{ij}M_{ij}\right)} \\ \dfrac{1}{2\Delta^2}\dfrac{L_{ij}^T M_{ij}}{M_{ij}M_{ij}} \leq 0 \quad \text{then} \quad C_T = 0 \end{cases} \tag{14}$$

Therefore, another problem of this model is the inaccurate results near the solid wall (e.g. near of terrain) because in this region the viscous layer is present, and the sub-grid stress is a significant fraction of the total stress. These models are incapable of computing the small scales present in this region [21] and lead to the underestimation of coefficient C_T, resulting in an overestimation of the velocity flow near the solid wall [22]. This has been accomplished by various types of *ad hoc* corrections such as *Van Driest damping* [23] and intermittence functions [24]. The *Van Driest damping* functions applied to the first row of equation (14) yield

$$C_T = \frac{1}{2\Delta^2}\frac{L_{ij}^T M_{ij}}{M_{ij}M_{ij}}\underbrace{\left[1-e^{-z^+/25}\right]^2}_{\text{Van Driest damping}} \quad \text{when} \quad C_T > 0 \tag{15}$$

where $z^+ = {}^{zu^*}/_{\nu}$ denotes the vertical coordinate in wall units and u^* the friction velocity of flow.

The SGS stress tensor to scalar variables is computed in some way:

$$\tau_{i\psi} = -\frac{\nu_T}{\Pr}\frac{\partial \psi^{\oplus}}{\partial x_i} = -\frac{C\Delta^2}{\Pr}\left|S^{\oplus}\right|\frac{\partial \psi^{\oplus}}{\partial x_i} \tag{16}$$

where Pr is the *Prandtl number,* a variable computed dynamically. The *Germano identity* applied to scalar ψ:

$$Q_i^T = \Im_{i\psi} - \tau_{i\psi}^T = \left(u_i^{\oplus}\psi^{\oplus}\right)^T - u_i^{\oplus T}\psi^{\oplus T} \tag{17}$$

where $I_{i\psi}$ denotes the SGS stress tensor according to the test filter. This tensor can be modelled as

$$\Im_{i\psi} = -\frac{\nu_T^T}{\Pr}\frac{\partial \psi^{\oplus T}}{\partial x_i} = -\frac{C_T(\alpha\Delta)^2}{\Pr}\left|S^{\oplus}\right|^T\frac{\partial \psi^{\oplus T}}{\partial x_i} \tag{18}$$

The relationship in Eq. (17) can be estimated by Eqs. (16) and (18):

$$\begin{aligned} Q_i^T &= -\frac{C_T\Delta^2}{\Pr}\underbrace{\left[\alpha^2\left|S^{\oplus}\right|^T\frac{\partial \psi^{\oplus T}}{\partial x_i} - \left(\left|S^{\oplus}\right|\frac{\partial \psi^{\oplus}}{\partial x_i}\right)^T\right]}_{-N_i} \\ &= \frac{C_T\Delta^2}{\Pr}N_i \end{aligned} \tag{19}$$

The least squares minimization of error function $\varepsilon rr = \left(Q_i^T - \frac{C_T\Delta^2}{\Pr}N_i\right)^2$ yield

$$\frac{1}{\Pr} = \frac{1}{C_T\Delta^2}\frac{Q_i^T N_i}{N_i N_i} \tag{20}$$

The computation of the SGS stress tensor ν_T in the eddy viscosity models (Eq. (8)) and the *Prandtl number* Pr in the turbulent diffusivity model (Eq. (16)) has been introduced into the ARPS code by Aguirre [8].

2.3. The parametrization of microphysical processes

In addition, the ARPS code has been designed specifically to describe thunderstorms. So, sub-models of heat flow and water steam, cloud formation and precipitation are included. Ground relief, vegetation types, soil types and initial conditions of the atmospheric state are taken into account for the simulation of these phenomena. The parameterization model is called warm-rain microphysics and is based on the descriptions of Klemp [25] and Soong [26]. This model considers three water categories such as water vapour (*qv*), cloud water (*qc*) and rainwater (*qr*), where the latter two are characterized by its size. At the beginning of the process, cloud water droplets are formed when the air becomes saturated and condensation occurs. Then, if the water mixing ratio exceeds a critical threshold in the cloud interior, raindrops form and the collision-coalescence process begins. If after crossing the base of the cloud they meet air below the saturation point, then the process of evaporation occurs, gradually diminishing the size of the raindrops. The time rate of evaporation and condensation will depend on certain parameters and on the water mixing ratio in the air. As all these processes involve transference of energy in heat and latent vapour, it is necessary to fit the potential temperature of the air. The theoretical formulation will not be described here because it is not the aim of this chapter. For more details on the microphysical processes in the ARPS code, see Xue [27].

2.4. Stochastic Lagrangian one-fluid particle model (STO)

The SGS models simulate the effect of unsolved scales as an energy sink of the solved LES scales. However, atmospheric dispersion phenomena are well simulated when small unsolved scales trajectory are computed. To obtain a more realistic description of the trajectories of fluid particles that transport the chemical species in a turbulent regime, the small-scale simulation unsolved by LES must be carried out. A coupled model between the LES and *stochastic Lagrangian models* (STO) for the small scales is proposed. The fluid particle *Lagrangian* velocity U_i is computed by solving the LES-STO coupled model by using a *Langevin* equation:

$$\frac{dU_i}{dt} = h_{ij}(U_j,t) + q_{ij}(U_j,t)\eta_j(t) \tag{21}$$

where h_{ij} is the dynamic deterministic coefficient, and q_{ij} is the dynamic random coefficient (in analogy with the Brownian movement), and it is linked to the statistical properties of the turbulence. η_j denotes a random variable whose mean value is null and covariance:

$$\langle \eta_i(t')\eta_j(t'')\rangle = \delta_{ij}\delta(t'-t'') \tag{22}$$

http://dx.doi.org/10.5772/63552

This property suggests that η_j is correlated neither in space nor in time.

To obtain the *Langevin* equation terms, the following hypotheses are proposed:

a. The small scales of the turbulence are statistically isotropic; that is, they lose the memory of eddy geometry that originated them.

b. They are located far from the *Kolmogorov* inertial range energy spectrum, then away from the production area of kinetic energy.

c. There is transfer of energy from large scales towards small ones, which is dissipated by molecular viscosity.

The first is not completely sustainable in closeness to solid walls or to the ground. For this reason, an anisotropic model will be used. The last hypothesis suggests that a relationship exists between the results of the LES and the coefficients of the *Langevin* model of the equation (21). In this way, these terms are calculated dynamically in each cell and for each time step from the results of the LES. Consequently, the fluid particle moves inside the cell of the calculation grid by following the evolution of the large scales resolved by LES to which is added a fluctuation that simulates the behaviour of the small scales of movement produced in its interior. The fluid particle velocity decomposition can be expressed as

$$U_i(x_i,t) = u_i^{\oplus}(x_i,t) + u_i^{-}(x_i,t) \tag{23}$$

where u_i^{-} is a *Lagrangian* fluctuation velocity of fluid particle due to sub-grid scale (SGS) turbulence. To obtain the coefficients of Eq. (21), the transport equation of probability density function (PDF) applied to velocities field, called the *Fokker-Plank* equation, can be used [4]:

$$\frac{\partial P_L}{\partial t} = -\frac{\partial}{\partial v_i}\left[h_{ij}(U_j,t)P_L\right] + \frac{1}{2}\frac{\partial^2}{\partial v_i \partial v_j}\left[q_{ik}(U_i,t)q_{jk}(U_j,t)P_L\right] \tag{24}$$

The decomposition Eq. (23) is used into the PDF transport Eq. (24):

$$\begin{aligned}\frac{\partial P_L}{\partial t} = &-\frac{\partial}{\partial v_i}\left[h_{ij}^{\oplus}(u_j^{\oplus},t)P_L\right] - \frac{\partial}{\partial v_i}\left[h_{ij}^{-}(u_j^{-},t)P_L\right] \\ &+\frac{1}{2}\frac{\partial^2}{\partial v_i \partial v_j}\left[q_{ik}^{\oplus}(u_i^{\oplus},t)q_{jk}^{\oplus}(u_j^{\oplus},t)P_L\right] + \frac{1}{2}\frac{\partial^2}{\partial v_i \partial v_j}\left[q_{ik}^{-}(u_i^{-},t)q_{jk}^{-}(u_j^{-},t)P_L\right]\end{aligned} \tag{25}$$

Also the same decomposition Eq. (23) is used in the *Langevin* Eq. (21):

$$\frac{du_i^{\oplus}}{dt} + \frac{du_i^{-}}{dt} = h_{ij}^{\oplus}(u_j^{\oplus},t) + h_{ij}^{-}(u_j^{-},t) + \left[q_{ij}^{\oplus}(u_j^{\oplus},t) + q_{ij}^{-}(u_j^{-},t)\right]\eta_j(t) \tag{26}$$

In this analysis, some simplifications in Eq. (26) with reference to the hypothesis mentioned above can be performed. In particular, the hypothesis (b) proposes that small scales are far from the inertial range energy transfer, then it can be assumed that the random term $q_{ij}$is completely defined by u_j^- small scales (sub-grid scales); in other words $q_{ij}^{\oplus}(u_j^{\oplus}, t)=0$. Therefore, separating large scales and sub-grid scales of Eq. (26) can be assigned the following equivalences:

$$\begin{cases} \dfrac{du_i^{\oplus}}{dt} = h_{ij}^{\oplus}(u_j^{\oplus},t) \\ \dfrac{du_i^-}{dt} = h_{ij}^-(u_j^-,t) + q_{ij}^-(u_j^-,t)\eta_j(t) \end{cases} \tag{27}$$

The first row of Eq. (27) is the material derivative of large scales. It is computed according to Eq. (4):

$$\frac{du_i^{\oplus}}{dt} = g_i B^{\oplus} - \frac{1}{\rho}\frac{\partial p'^{\oplus}}{\partial x_i} - \frac{\partial \tau_{ij}}{\partial x_j} + 2\nu \frac{\partial S_{ij}^{a\oplus}}{\partial x_j} \tag{28}$$

Therefore, the first term of the second member of Eq. (25) is obtained from LES:

$$-\frac{\partial}{\partial v_i}\left[h_{ij}^{\oplus}(u_j^{\oplus},t)P_L\right] = -g_i B^{\oplus}\frac{\partial P_L}{\partial v_i} + \frac{1}{\rho}\frac{\partial p'^{\oplus}}{\partial x_i}\frac{\partial P_L}{\partial v_i} + \frac{\partial \tau_{ij}}{\partial x_j}\frac{\partial P_L}{\partial v_i} - 2\nu\frac{\partial S_{ij}^{a\oplus}}{\partial x_j}\frac{\partial P_L}{\partial v_i} \tag{29}$$

For the other terms of Eq. (25), a turbulence model is proposed because these are terms dependent on the unsolved scales. For deterministic term, which takes account of unsolved scales, Gicquel [7] suggested that this value is proportional at velocity of small scales and a tensor that note the turbulent kinetic energy of sub-grid scales. The idea of this proposal is that the greater the turbulent kinetic energy of sub-grid scales, the greater is the number of small vortices and therefore the greater importance of this term. The other terms express the random component sub-grid. They can be modelled taking into account the assumption (a); in other words, the small scales have an isotropic behaviour. Pope [6] proposed an expression for these terms which takes into account also the rate of dissipation in homogeneous isotropic sub-grid turbulence ε.

With these considerations, the transport equation for the probability density function of velocity field in Eq. (25) can be summarized as

$$\frac{\partial P_L}{\partial t} = -\frac{\partial}{\partial v_i}\left[\frac{du_j^{\oplus}}{dt}P_L\right] - \frac{\partial}{\partial v_i}\left[\alpha_{ij}u_j^- P_L\right] + \frac{1}{2}C_0\varepsilon\frac{\partial^2 P_L}{\partial v_i \partial v_j} \tag{30}$$

where $C_0=2.1$ denotes the *Kolmogorov coefficient*.

With the hypotheses and using the PDF transport Eq. (30), the dynamic deterministic coefficient and dynamic random coefficient of Eq. (21) can be expressed as

$$\begin{cases} h_{ij}(U_j,t) = \dfrac{du_j^{\oplus}}{dt} + \alpha_{ij} u_{ij}^{-} \\ q_{ij}(U_j,t) = \sqrt{C_0 \varepsilon}\ \delta_{ij} \end{cases} \tag{31}$$

Since the material derivative of large scales is obtained using LES in Eq. (28) and the rate of dissipation of turbulent kinetic energy ε is computed with a gradient model [28] by the ARPS code, the remainder proposes an expression for the deterministic tensor α_{ij}. These tensors are related to the statistical properties of sub-grid turbulence [29]. For the more complex case, these features are not steady, inhomogeneous and anisotropic flow, as developed in the atmospheric boundary layer over heterogeneous rough terrain near the soil surface:

$$\alpha_{ij} = \frac{1}{2K^-}\frac{dK^-}{dt}\delta_{ij} - \left(\frac{3}{4}C_0\right)\frac{\varepsilon}{K^-}\delta_{ij} + \left(\frac{R_{ij}}{2K^-} - \frac{\delta_{ij}}{3}\right)\frac{\varepsilon}{K^-} \tag{32}$$

where $K^- = \frac{1}{2}\left(u_i^- u_j^-\right)^{\oplus}\delta_{ij}$ denotes the sub-grid turbulent kinetic energy; it is obtained with transport Eq. (28). $R_{ij} = \left(u_i^- u_j^-\right)^{\oplus}$ denotes the Reynolds tensor of sub-grid scales; it is obtained using the turbulent diffusivity models (Eq. 8). More details of this resolution can be found in [9, 10] and [29].

2.5. Solid particle simulation model

The solid particle simulation addressed in this work is on the order of tens of micron size. Therefore, the following assumptions may be raised:

a. The forces due to gravity and viscosity are dominant regarding the other forces on the solid particles.

b. The shape of the solid particles is approximately spherical.

c. The concentration of solid particles is not sufficiently large to influence kinematic or thermodynamic properties of air transports.

The motion equation of a solid particle immersed in fluid flow can be written as follows [30]:

$$\frac{dV_i}{dt} = F_i \tag{33}$$

where V_i is the solid particle velocity, and F_i denotes the forces per unit mass.

The assumption (a) has been proven when the density of the solid particles is more than 1000 times of the air density which transports. Then, the second member (Eq. (33)) can be summarized as

$$F_i = \frac{U_i - V_i}{\tau_s} - g_i \delta_{i3} \tag{34}$$

in which the first term on the right-hand side denotes the drag force, while the second term is the force of gravity acting in the vertical direction indicated with the index 3, both per unit mass of solid particle. τ_s denotes a time scale of particle acceleration. It depends on the rate of densities between the solid particle and the air, solid particle diameter *ds* and a drag coefficient C_D.

$$\tau_s = \frac{4}{3}\frac{ds}{C_D}\frac{\rho_s}{\rho}\frac{1}{|U_i - V_i|} \tag{35}$$

The hypothesis (b) suggests that the drag coefficient C_D is proportional to the diameter of the solid particle *ds* and also depends on the fluid viscosity. This condition is usually assessed by calculating the *Reynolds number of the solid particle* $\Re e_s$.

$$\Re e_s = d_s \frac{|U_i - V_i|}{\nu} \tag{36}$$

where ν denotes the dynamic viscosity of fluid.

In laminar flow (or *Stokes flow*), $\Re e_s \leq 1$ and the drag coefficient can be calculated as

$$C_D = \frac{24}{\Re e_s} \tag{37}$$

In turbulent flows, the authors propose many different expressions. Sommerfeld [31] uses the following:

$$C_D = \begin{cases} \frac{24}{\Re e_s}\left(1 + 0.15\Re e_s^{0.687}\right) & \text{if} \quad 1 < \Re e_s < 1000 \\ 0.44 & \text{if} \quad \Re e_s \geq 1000 \end{cases} \tag{38}$$

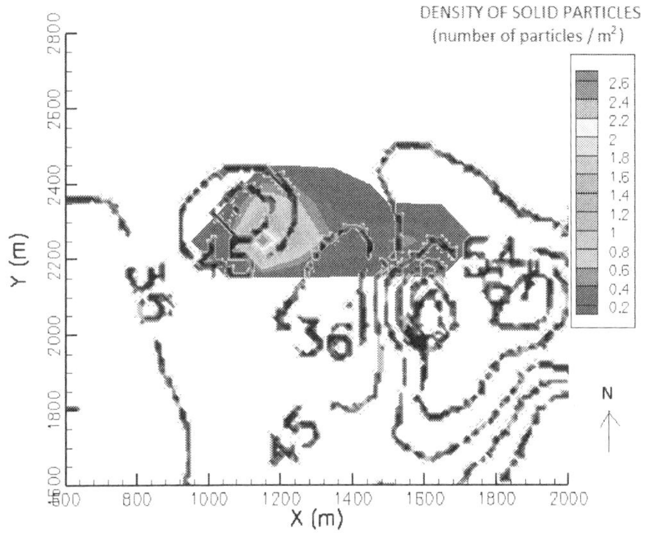

Figure 10. Copper concentration at ground level. Solid colour: simulated with 100 × 100 m grid mesh (part/m^2). Lines: measurements (mg_{Cu}/kg_{soil}) with 100 × 100 m by (Fernández-Turiel [14]).

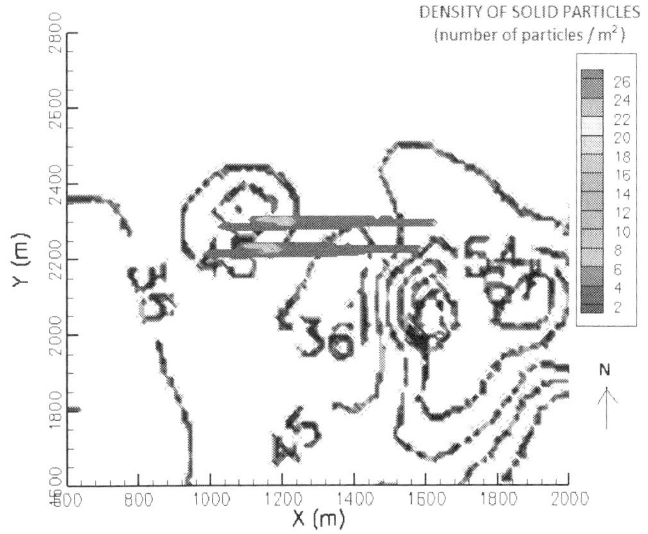

Figure 11. Copper concentration at ground level. Solid colour: simulated with 10 × 10 m grid mesh (part/m^2). Lines: measurements (mg_{Cu}/kg_{soil}) with 100 × 100 m by (Fernández-Turiel [14]).

Figure 12. Copper concentration at ground level simulated with 10 × 10 m grid mesh on a satellite image.

4.4. Discussion

In **Figure 10**, it is noted that concentrations simulated by the model are calculated by counting the number of copper particles falling in grids of 100 × 100 m similar to those samples that were used in Fernández-Turiel [14]. Concentration contours indicate the soil contamination that has been produced by the smelter during operation; simulating this only considers pollution with the eastern wind. Considering this situation, it can be seen that there is good agreement between the maximum value that indicates the model with closed contour 45 mg_{Cu}/Kg_{soil} concentration. Other major peaks of concentration are studied by Fernández-Turiel [14], but they are associated with other wind directions. However, when the concentrations simulated by the model are calculated using grids 10 × 10 m, as shown in **Figure 11**, they can be individualized with the plumes of each chimney.

The model shows that concentrations of copper particles have little lateral dispersion. This implies that if concentrations are calculated using smaller grid cells, they will be much higher in the area where they fall. This is the reason why both figures legends differ by a factor of 10. It is likely that if the particle size was smaller, lateral dispersion increases.

The concentrations shown by Fernández-Turiel [14] to the west from the location of the smelter still indicate the existence of copper particles beyond those shown in the simulation. If the copper particles are simulated with different diameters, it is likely that smaller particles travel on greater distance before reaching the ground.

http://dx.doi.org/10.5772/63552

5. Conclusions

We have emphasized the use of computer simulation tools for the atmospheric dispersion of gases and solid particles in this chapter.

The first part of the chapter presented the large eddy simulation (LES) approach in order to numerically solve the turbulent flows of the great Reynolds number as those presented in the atmospheric boundary layer. Next, the coupling between the large eddy simulation (LES) and Lagrangian Stochastic one particle model (STO) was presented to detail how to dynamically calculate the model coefficients based on the kinetic energy of turbulent fluid flow.

The validation of these tools has relied on experimental measurements in a wind tunnel and in situ measurements. The stochastic Lagrangian one-particle method enables simulations without much computational cost and with good results. The results of the numerical simulation using the ARPS code (LES-STO) of the concentrations of CO_2 emitted by a chimney up wind uphill are in good agreement with experimental measurements performed in a wind tunnel [13].

Fernández-Turiel [14] has found high levels of trace of copper in soil and plants in the vicinity of smelter. As a consequence, it is important to determine the extent of the contaminated area and the concentration of these elements that might be potential hazards due to inhalation and ingestion. The model has been used to simulate the atmospheric dispersion of these particles emitted from the smelter chimneys during a severe eastern wind event. The results closely matched field measurements made by these authors.

Finally, this approach has been able to predict the dispersion and level concentration of pollutants (gases and/or solid particles) in the atmosphere with the aim of predicting the impact on the populations and preventing environments problems. This information can be used to determine the affected areas close to industrial factories and emission control protocols that need to be used during specific weather conditions. By understanding the local pollution concentrations and their movement, future industrial factories sites could be installed in locations that would minimize impacts to the surrounding population.

Author details

César Augusto Aguirre[1,2*] and Armando Benito Brizuela[2]

*Address all correspondence to: cesaraguirredalotto@gmail.com

1 Center of Scientific Research and Technology Transfer Production - National Scientific and Technical Research Council - Argentina (CICyTTP - CONICET), Diamante, Entre Rios, Argentina

2 School of Agricultural Sciences - National University of Entre Ríos (FCA - UNER), Oro Verde, Entre Rios, Argentina

References

[1] Deardorff J. W. *A numerical study of three-dimensional turbulent channel flow*. Journal of Fluid Mechanics. 1970; 41:453–480.

[2] Deardorff J. W. *The use of sub-grid transport equations in a three-dimensional model of atmospheric turbulence*. Journal of Fluid Engineering. 1973; 4:429–438.

[3] Schumann U. *Sub-grid scale model for finite difference simulations of turbulent flow in plane channels and annuli*. Journal of Computational Physics. 1975;18:376–404.

[4] Pope S. B. PDF methods for turbulent reactive flows. Energy Combustion Science. 1985;11:119–192.

[5] Haworth D. C. and Pope S. B. *A generalized Langevin model for turbulent flow*. Physics of Fluids. 1986;29(2):378–405.

[6] Pope S. B. *Lagrangian PDF methods for turbulent flows*. Annual Review of Fluid Mechanics. 1994;26:23–36.

[7] Gicquel L. Y. M., Givi P., Jaberi F. A. and Pope S. B. *Velocity filtered density function for large-eddy simulation of turbulent flow*. Physics of Fluids. 2002;14(3):1196–1213.

[8] Aguirre C. A. *Dispersión et Mélange Atmosphérique Eulèro-lagrangien de Particules Fluides Réactives. Application à des cas simples et complexes* [thesis]. Université Claude Bernard, Lyon, France: Ecole Doctorale MEGA; 2005. 323 p. Available from: http://venus.ceride.gov.ar/twiki/pub/Cimec/RepositorioDeInformesTesis/aguirre-phd-thesis.pdf

[9] Aguirre C. A., Brizuela A. B., Vinkovic I. and Simoëns S. *A sub-grid Lagrangian stochastic model for turbulent passive and reactive scalar dispersion*. International Journal of Heat and Fluid Flow. 2006;27(4):627–635.

[10] Aguirre C. A., Brizuela A. B., Vinkovic I. and Simoëns S. Eulero-Lagrangian coupled model for the Simulation of atmospheric dispersion of chemically reactive species into the boundary layer. Serie Mecánica Computacional. 2006, 25 (2): 185-205.

[11] Vinkovic I., Aguirre C. A., Ayrault M. and Simoëns S. *Large-eddy simulation of the dispersion of solid particles in a turbulent boundary layers*. Journal of Boundary Layers Meteorology. 2006;121:283–311. DOI: 10.1007/s10546-006-9072-6

[12] Vinkovic I., Aguirre C. A., Simoëns S. and Gorokhovski M. *Large-eddy simulation of droplet dispersion for inhomogeneous turbulent wall flow*. International Journal of Multiphase Flow. 2006;32(3):344–364.

[13] Gong W. *A wind tunnel study of turbulent dispersion over two – and three- dimensional gentle hills from upwind point sources in neutral flow*. Journal of Boundary Layer Meteorology. 1991;54:211–230.

[14] Fernández-Turiel J. L., Aceñolaza P., Medina M. D., Llorens J. F. and Sardi F. *Assessment of a smelter impact area using surface soils and plants*. Journal of Environmental Geochemistry and Health. 2001;23:65–78.

[15] Leonard A. *Energy cascade in large eddy simulation of turbulent fluid flow*. Advances in Geophysics. 1974;18(A):237–248.

[16] Smagorinsky J. *General circulation experiments with the primitive equations. I. The basic experiments*. Monthly Weather Review. 1963;91:99–164.

[17] Lilly D. K. A. *Proposed modification of the Germano subgrid-scale closure method*. Physics of Fluids. 1992;4(A):633–635.

[18] Germano M., Piomelli U., Moin P. and Cabot W. H. *A dynamic subgrid-scale eddy viscosity model*. Physics of Fluids. 1991;A(3):1760–1765.

[19] Lévêque E., Toschi F., Shao L. and Bertoglio J-P. *Shear-improved Smagorinsky model for large-eddy simulation of wall-bounded turbulent flows*. Journal of Fluid Mechanics. 2007;570:491–502.

[20] Piomelli U. and Balaras E. *Wall-layer models for large-eddy simulations*. Annual Review of Fluid Mechanics. 2002;34:349–374.

[21] Porté-Agel F., Meneveau C. and Parlange M. *A scale-dependent dynamic model for large-eddy simulation: application to a neutral atmospheric boundary layer*. Journal of Fluid Mechanics. 2000;415:261–284.

[22] Cui G. X., Xu C. X., Fang L., Shao L. and Zhang Z. S. *A new subgrid eddy-viscosity model for large-eddy simulation of anisotropic turbulence*. Journal of Fluid Mechanics. 2007;582:377–397.

[23] Van Drieas E. R. *On turbulent flow near wall*. Journal of Aero Sciences. 1956;23:1007–1011.

[24] Piomelli U., Ferziger J., Moin P. and Kim J. *New approximate boundary conditions for large-eddy simulation of wall-bounded flows*. Physics of Fluids. 1989;A(1):1061–1068.

[25] Klemp J. B. and Wilhelmson R. B. *The simulation of three-dimensional convective storm dynamics*. Journal of Atmospheric Science. 1978;35:1070–1096.

[26] Soong S-T. and Oruga Y. *A comparison between axi-symmetric and slab-symmetric cumulus cloud models*. Journal of Atmospheric Science. 1973;30:879–893.

[27] Xue M., Drogemeier K. and Wong V. *The Advanced Regional Prediction System (ARPS). A multi-scale non-hydrostatic atmospheric simulation and prediction model. Part I: model dynamics and verification*. Meteorology Atmospheric Physics. 2000;75:161–193.

[28] Deardorff J. W. *Stratocumulus-capped mixed layer derived from a three dimensional model*. Boundary Layer Meteorology. 1980;18:495–527.

[29] Aguirre C. A. and Brizuela A. B. Numerical Simulation of Atmospheric Dispersion Gas Passive on a hill using a coupled model. Serie Mecánica Computacional. 2008; 27 (4): 217-237.

[30] Kosinski P., Kosinska A. and Hoffmann A. C. *Simulation of solid particles behaviour in a riven cavity flow*. Powder Technology. 2009;191:327–339.

[31] Sommerfeld M. *Analysis of collision effects for turbulent gas–particle flow in a horizontal channel: Part I. Particle transport*. International Journal of Multiphase flow. 2003;29:675–699.

[32] Aguirre C. A., Y. Guo and Ayrault M. Dispersion of solid particles in saltation movement in a turbulent flow. Journal Comptes Rendus Mécanique. 2004; 332: 627-632.

[33] Aguirre C. A., Simoëns S. and Ayrault M. *Dispersed of solid heavy particles in a homogeneous turbulence*. In: Brebbia C. A. and Martin Duque J. F., editors. *Air Pollution*. 10th ed. Southampton, UK: Wessex Institute of Technology Press; 2002. p. 6.

[34] Gong W. and Ibbetson A. *A wind tunnel study of turbulent flow over models hill*. Boundary Layer Meteorology. 1989;49:113–148.

[35] Counihan J. *An improved method of simulating an atmospheric boundary layers in a wind tunnel*. Journal of Atmospheric Environment. 1969;3:197–214.

[36] Justus C. G. *Wind and system performance*. Philadelphia, Penn, USA: The Franklin Institute Press; 1978. 120 p.